Lecture Notes in Computer Science 14939

Founding Editors

Gerhard Goos
Juris Hartmanis

The series Lecture Notes in Computer Science (LNCS), including its subseries Lecture Notes in Artificial Intelligence (LNAI) and Lecture Notes in Bioinformatics (LNBI), has established itself as a medium for the publication of new developments in computer science and information technology research, teaching, and education.

LNCS enjoys close cooperation with the computer science R & D community, the series counts many renowned academics among its volume editors and paper authors, and collaborates with prestigious societies. Its mission is to serve this international community by providing an invaluable service, mainly focused on the publication of conference and workshop proceedings and postproceedings. LNCS commenced publication in 1973.

Emil Sekerinski · Leila Ribeiro

Editors

Formal Methods Teaching

6th Formal Methods Teaching Workshop, FMTea 2024
Milan, Italy, September 10, 2024
Proceedings

 Springer

Editors
Emil Sekerinski 🆔
McMaster University
Hamilton, ON, Canada

Leila Ribeiro 🆔
Universidade Federal do Rio Grande do Sul
Porto Alegre, Brazil

ISSN 0302-9743   ISSN 1611-3349 (electronic)
Lecture Notes in Computer Science
ISBN 978-3-031-71378-1   ISBN 978-3-031-71379-8 (eBook)
https://doi.org/10.1007/978-3-031-71379-8

Preface

Formal methods of programming were conceived starting in the 1970s to explain systematic program development to novice programmers. They were also seen as necessary for a mature engineering discipline. Today, formal methods thrive in niche areas in the industry, e.g. where safety and security are of existential importance, as witnessed, for example, by the FM 2024 Industry Day. These applications of formal methods are driven by highly dedicated and educated practitioners. However, while everyday software development would benefit equally from formal methods, the post-secondary education of programmers has not met that goal. Many questions remain: Which formalisms should be taught in dedicated courses, and which should support existing courses? Which formal tools should be taught early in an undergraduate curriculum, and which at the graduate level? Is it more pedagogical to integrate or not integrate formal methods and their associated tools with existing languages and tools? How can students embrace learning formal methods if employers do not ask for them? What are the principles and tools for teaching and grading large classes with formal methods? The FMTea Workshop Series (https://fmtea.github.io), organized by FME's Teaching Committee (https://fme-teaching.github.io), seeks to answer these and related questions.

The FMTea 2024 Workshop took place on September 10, 2024, in Milan, Italy, as part of the 26th International Symposium on Formal Methods. Of nine submissions, each reviewed by three Program Committee Members in a single-blind process, seven were selected for presentation and inclusion in these proceedings. André Platzer, Karlsruhe Institute of Technology, Germany, kindly agreed to contribute with a paper and talk on *The Significance of Symbolic Logic for Scientific Education.* New in this edition of FMTea is a Formal Methods Teaching Expo to foster the exchange of teaching experiences and the adoption of teaching material. The presentations of the Expo are not included in these proceedings.

We wish to express our gratitude to the FME Teaching Committee and, in particular, its Chair, Luigia Petre, for guidance, to the Program Committee Members for their dedication, to the EasyChair organization for their technical infrastructure, to Springer for publishing the proceedings in the Formal Methods Teaching Series (https://link.springer.com/conference/tfm), and to the General Chairs of FM 2024, Matteo Pradella and Matteo Rossi, and the Workshop Chairs, Stefania Gnesi and Marieke Huisman, for their organizational support.

July 2024

Emil Sekerinski
Leila Ribeiro

Organization

Program Committee Chairs

Emil Sekerinski   McMaster University, Canada
Leila Ribeiro   Universidade Federal do Rio Grande do Sul, Brazil

Program Committee

Erika Ábrahám	RWTH Aachen University, Germany
Sandrine Blazy	University of Rennes 1, France
Brijesh Dongol	University of Surrey, UK
Catherine Dubois	ENSIIE, France
João F. Ferreira	INESC-ID & IST, University of Lisbon, Portugal
Markus Kuppe	Microsoft, USA
Thierry Lecomte	CLEARSY, France
Michael Leuschel	University of Düsseldorf, Germany
Alexandra Mendes	INESC TEC & Faculty of Engineering, University of Porto, Portugal
Tim Nelson	Brown University, USA
David Pearce	ConsenSys & Victoria University of Wellington, New Zealand
Luigia Petre	Åbo Akademi University, Finland
Pierluigi San Pietro	Politecnico di Milano, Italy
Graeme Smith	University of Queensland, Australia

Contents

Invited Talk

The Significance of Symbolic Logic for Scientific Education

André Platzer[(✉)]

Karlsruhe Institute of Technology, Karlsruhe, Germany
platzer@kit.edu

Abstract. This invited paper is a passionate pitch for the significance of logic in scientific education. Logic helps focus on the essential core to identify the foundations of ideas and provides corresponding longevity with the resulting approach to new and old problems. Logic operates symbolically, where each part has a precise meaning and the meaning of the whole is compositional, so a simple function of the meaning of the pieces. This compositionality in the meaning of logical operators is the basis for compositionality in reasoning about logical operators. Both semantic and deductive compositionalities help explain what happens in reasoning. The correctness-critical core of an idea or an algorithm is often expressible eloquently and particularly concisely in logic. The opinions voiced in this paper are influenced by the author's teaching of courses on cyber-physical systems, constructive logic, compiler design, programming language semantics, and imperative programming principles. In each of those courses, different aspects of logic come up for different purposes to elucidate significant ideas particularly clearly. While there is a bias of the thoughts in this paper toward computer science, some courses have been heavily frequented by students from other majors so that some transfer of the thoughts to other science and engineering disciplines is plausible.

Keywords: Education · Logic · Logic of dynamical systems ·
Constructive logic · Proofs · Programs · Program semantics

1 Introduction

This paper is a passionate pitch for the significance of logic in scientific education. Education has multiple important goals that range from practical skills that enable a student to efficiently solve present challenges met in academic and industrial practice all the way to the longevity of foundations that influence a student's thinking for a lifetime in yet unforeseeable areas. Logic is particularly good at impacting the latter foundations but also plays a role in the former practice. The reason for logic's longevity is its unreasonable effectiveness [18] of identifying the core essentials of a question and its answer. For example, several aspects that make a big syntactic difference in a particular application context

Funding has been provided by an Alexander von Humboldt Professorship.

E. Sekerinski and L. Ribeiro (Eds.): FMTea 2024, LNCS 14939, pp. 3–22, 2024.
https://doi.org/10.1007/978-3-031-71379-8_1

still lead to a negligible difference once the phenomenon has been captured with logic. Just like proof-based mathematics, its sibling of logic teaches rigorous reasoning principles and how to use it to overcome challenges. Logic is crystal clear on the foundations of reasoning, on what a proof is, which reasoning principles are correct and why, and is explicit about possible structuring principles about proofs. There is an obvious and frequently mentioned positive synergy with experience in proof-based mathematics and experience in logic.

The opinions reflected in this paper are based on the author's experience in computer science undergraduate to PhD-level teaching at different research universities since 2008 as well as engagement in university committees restructuring undergraduate introductory education. While these opinions are influenced from the author's educational computer science bias, the introductory Principles of Imperative Computation and the upper-level Logical Foundations of Cyber-Physical Systems courses he taught were also frequented by students from several other majors, including mathematics, physics, electrical engineering, and robotics, so that some generalization of these thoughts to other sciences and engineering disciplines are plausible.

2   Logic in the Science Curriculum

To give the reader some appreciations for the different aspects of logic that enter different courses, here is an overview of some of the logic-influences in some of the courses influenced by what the author taught.[1]

1. Principles of Imperative Computation is the first computer science course for computer science undergraduates and many other disciplines originally designed by Frank Pfenning and refined by the author. It combines algorithms and data structures, imperative programming introduction, and program contracts for establishing their correctness. Logic enters informally when reasoning mathematically about preconditions, postconditions, loop invariants, and data structure invariants. This immediate appreciation for the fundamental aspects of operational and logical reasoning about programs enables the students in the course to obtain a particularly strong command of writing reliable imperative programs. In the last homework, the students showcase their understanding of imperative programming by programming a virtual machine in C reflecting on the nuances of operational semantics. About 300–800 students take this course every year.
2. Constructive Logic is a third year undergraduate course for computer science but also mathematics and philosophy students designed by Frank Pfenning, also taught by Karl Crary, and refined by the author. It teaches the logical foundations of functional programming from proofs-as-programs via the Curry-Howard isomorphism as well as the logical foundations of logic programming from propositions-as-programs while proving the fundamental principle of cut elimination and proof search refinements along the way. About 100 students take this course every year.

---

[1] Lecture material is available at http://lfcps.org/.

3. Compiler Design is an upper undergraduate level computer science course designed by Frank Pfenning and the author. It teaches the principles for designing compilers, where logical proof rules that represent logic programs are used to capture dataflow analysis, parsing, programming language semantics (via structural operational semantics), compiler optimizations etc. While logic is not a learning goal of the course, logic is still crucial in concisely communicating the actual idea behind many phases of a compiler. About 80 students take this course every year.

4. Logical Foundations of Cyber-physical Systems [12,15] is an upper undergraduate and/or master's level computer science course designed by the author. It teaches what principles one has to know in order to design safe cyber-physical systems and justify their safety properties. Logic and programming languages play central roles in the course as means to identify the core mathematical challenges and solution techniques especially for justifying safety. The course is supported by a textbook, slides, lecture videos, a theorem prover, active learning quizzes, and a competition. About 20 students take this course every year.

5. Programming Language Semantics is a PhD-level course based on Steve Brooke's take on John Reynolds course design [17], but reinterpreted by the author with more focus on logic. It teaches the principles of programming language semantics, which has a direct link to logic via the axiomatic semantics of programming languages, as well as clear links with logic via soundness and relative completeness theorems of the axiomatic semantics with respect to the denotational or operational semantics. The extensive use of logic in the author's version of the course significantly simplifies otherwise quite demanding and technical semantical challenges making it possible to teach crucial ideas with minimal technical effort. About 10–20 students take this course every year.

Some of the insights behind some of these courses with a particular focus on logic will be reviewed in the sequel. The course in items 1 is a required first semester undergraduate course. The courses in items 2 and 4 are one of several different courses satisfying the undergraduate logic requirement. The course in item 3 is one of several different courses satisfying the systems requirement. The courses in items 4 and 5 are one of several different courses satisfying the programming language PhD requirement.

3   Course: Logical Foundations of Cyber-Physical Systems

The *Logical Foundations of Cyber-Physical Systems* (LFCPS) course[2] is accompanied by a textbook [15], slides, more than 20 h of lecture videos[3], and a thorough active learning quiz that enables students to practice and reinforce the learning goals of every lecture. The course was originally designed for upper

---

[2] http://www.lfcps.org/lfcps/.
[3] http://videos.lfcps.org/.

level undergraduate students [12] but has since been opened up to masters and PhD students. It has been designed and taught by the author at Carnegie Mellon University, ENS Lyon, the University of Braga, and the Karlsruhe Institute of Technology. Short versions of the course were the basis for several summer school lectures, including the summer schools in Marktoberdorf, on Automated Reasoning, on Verification Technology Systems & Applications, and on Cyber-Physical Systems. The following discussion is based on thoughts expressed in the LFCPS textbook [15], which also identifies lecture dependencies for courses based on different subsets of the lectures.

The Logical Foundations of Cyber-Physical Systems textbook and course are breaking with the myth that cyber-physical systems (CPSs) are too challenging to be taught at the undergraduate level. CPSs such as computer-controlled cars, airplanes or robots play an increasingly crucial role in our daily lives. They are systems that we bet our lives on, so they need to be safe. Getting CPSs safe, however, is an intellectual challenge, because of their intricate interactions of complex control software with their physical behavior. Who can design these notoriously challenging systems with the scrutiny that is required to make sure they can be used safely? How can students, scientists, and practitioners acquire the required background in a single course or a single textbook in a way that meets the demands on rigor required in safe CPS design? In the LFCPS course, students quickly advance from learning basic concepts underlying CPSs to being able to prove safety properties about complex CPS.

The Challenge. Teaching CPS-related topics is notoriously challenging, but also creates an opportunity to discover and explore other areas of science with the intrinsic motivation it takes to succeed. A few students may have a background either in engineering physical systems, or in some areas of formal methods, but almost never in both and, in fact, often in neither of the two. A sharp educational gap has also been confirmed across the board at the 2013 NSF workshop on CPS education [3]. This brings up the question of how to best teach the core aspects of CPS with the rigor that is required to prepare students and professionals for the challenges that lie ahead in enriching our world with safe and reliable cyber-physical systems.

The challenge is that CPSs are a cozy topic to take on after background reading equivalent to the material acquired for a PhD in mathematics, a PhD in computer science, a PhD in logic, and a PhD in engineering or controls. The trick is to find out how to enable students to understand CPS gradually without losing interest and engagement. Besides identifying the easiest and most intuitive, background-free approach for presenting the various technical concepts, the biggest educational contribution of this course is the clever out-of-order arrangement of the topics to overcome the problem that CPSs have such a long dependency chain of required background. There certainly is a reason for the acute shortage of rigorous CPS courses for undergraduate students.

There are two primary ways of learning about cyber-physical systems [15], reprinted here with permission from Springer:

Onion Model. The *Onion Model* follows the natural dependencies of the layers of mathematics going outside in, peeling off one layer at a time, and progressing to the next layer when all prerequisites have been covered. This would require the CPS student to first study all relevant parts of computer science, mathematics, and engineering, and then return to CPS in the big finale. That would require the first part of the course to cover real analysis, the second part differential equations, the third part conventional discrete programming, the fourth part classical discrete logic, the fifth part theorem proving, and finally the last part cyber-physical systems. In addition to the significant learning perseverance that the Onion Model requires, a downside is that it misses out on the integrative effects of CPSs that can bring different areas of science and engineering together, and which provide a unifying motivation for studying them in the first place.

Scenic Tour Model. The LFCPS course follows the *Scenic Tour Model*, which starts at the heart of the matter, namely CPSs, going on scenic expeditions in various directions to explore the world around as we find the need to understand the respective subject matter. The course directly targets CPS right away, beginning with simpler layers that the reader can understand in full before moving on to the next challenge.

For example, the first layer comprises CPSs without feedback control, which allow simple finite open-loop controls to be designed, analyzed, and verified without the technical challenges considered in later layers of CPS. Likewise, the treatment of CPS is first limited to cases where the dynamics can still be solved in closed form, such as straight-line accelerated motion of Newtonian dynamics, before generalizing to systems with more challenging differential equations that can no longer be solved explicitly. This gradual development where each level is mastered and understood and practiced in full before moving to the next level is helpful to tame complexity. It also follows naturally the layers of complexity in logic. The Scenic Tour Model has the advantage that the students stay on CPSs the whole time, and leverage them as the guiding motivation for understanding more and more about the connected areas. It has the disadvantage that the resulting gradual development of CPS does not necessarily always present matters in the same way that an after-the-fact compendium would treat it.[4] A gradual development can also be more effective at conveying the ideas, reasons, and rationales behind the development compared to a final compendium, which improve generalizability promises compared to a mere factual presentation.

Computational Thinking for CPS. The approach that the LFCPS course follows takes advantage of Computational Thinking [19] for CPSs. Due to their subtleties and the intricate interactions of complex control software with the physical world, CPSs are notoriously challenging. Logical scrutiny, formalization, and thorough safety and correctness arguments are, thus, critical for CPS. Because CPS are so easy to get wrong, these logical aspects are an integral part of their design and critical to understanding their complexities.

---

[4] The textbook compensates for this by providing technical summaries and by highlighting important results for later reference.

The primary attention of the course, thus, is on the foundations and core principles of CPS. The course tames some of the CPS complexities by focusing on a simple core programming language for CPS. The elements of the programming language are introduced hand in hand with their reasoning principles, which makes it possible to combine CPS program design with their safety arguments. This is important, not just because abstraction is a key factor for success in CPS, but also because retrofitting safety is not possible in CPS. The CPS programming language of hybrid programs taught in the textbook pass with flying colors Alan J. Perlis' test, who rejected programming languages by the following criterion:

> "A language that doesn't affect the way you think about programming, is not worth knowing."
> – Alan J. Perlis [7]

Logic to Tame CPS. On account of their technical challenges, programming and investigating the safety of CPS may be a daunting task. But logic significantly simplifies this challenge thanks to the principles of logical compositionality in multi-dynamical systems [11,13,15]. The first step is to make safety statements about cyber-physical systems a first-level citizen in logic. The resulting differential dynamic logic (dL) [8–11,14,15] features modalities $[\alpha]$ for hybrid programs α that describe the possible behavior of a CPS as a program with differential equations. The dL formula $[\alpha]P$ expresses that after all runs of hybrid program α the dL formula P is true (safety). It is an ordinary dL formula, so the implication $Q \to [\alpha]P$ expresses that if formula Q is true initially, then all runs of hybrid program α are such that formula P is true afterwards.

Example 1 (Car acceleration or braking). The following dL formula expresses that, if a car x is before an obstacle m and its brakes b work, then all ways of following a hybrid program that first has a nondeterministic choice ($\cup$) to apply acceleration by assigning $a := A$ or to apply brake by $a := -b$ and subsequently follows the differential equation system with the time-derivative x' of position x being velocity v, whose time-derivative is the acceleration a but only while the velocity v is nonnegative, then all its behaviors keep the position before the obstacle and the velocity nonnegative:

$$x \leq m \wedge b > 0 \to [(a := A \cup a := -b); \{x' = v, v' = a \,\&\, v \geq 0\}](x \leq m \wedge 0 \leq v)$$

Whether this dL formula is true is a good question, but now this question has a logically precise rendition and unambiguous answer.

Logic is not just useful for the clear and unambiguous expression of questions about cyber-physical systems but also for finding the answer. This is where axioms for cyber-physical systems become useful. For example, the dL axiom [;] captures that all behavior of the sequence $\alpha; \beta$ safely satisfies formula P if and only if all behaviors of the first part α are such that all behaviors of the second part β satisfy P:

$$[;] \quad [\alpha; \beta]P \leftrightarrow [\alpha][\beta]P$$

Example 2 (Car motion after car control). Using axiom [;] reduces the dL formula from Example 1 to an equivalent that split off the discrete control from the differential equation of motion:

$$x \leq m \wedge b > 0 \rightarrow [a := A \cup a := -b][x' = v, v' = a \,\&\, v \geq 0](x \leq m \wedge 0 \leq v)$$

The advantage of this equivalent decomposition is that it separates the discrete control actions from the continuous motion such that both can be analyzed separately. Subsequent logical decompositions of the remaining parts in dL will ultimately lead to a significantly easier equivalent.

The following dL equivalence can be used to analyze or prove conjunctions in the safety conditions separately:

$$[]\wedge \ [\alpha](P \wedge Q) \leftrightarrow [\alpha]P \wedge [\alpha]Q$$

Example 3 (Car safety and speed separated). Using axiom []∧ reduces the dL formula from Example 2 to an equivalent that separately considers the position safety of $x \leq m$ and the speed safety $v \geq 0$:

$$x \leq m \wedge b > 0 \rightarrow \big([a := A \cup a := -b][x' = v, v' = a \,\&\, v \geq 0]\, x \leq m$$
$$\wedge \ [a := A \cup a := -b][x' = v, v' = a \,\&\, v \geq 0]\, 0 \leq v\big)$$

Indeed the last conjunct is fairly easy to establish, because the differential equation system $x' = v, v' = a \,\&\, v \geq 0$ is limited to $v \geq 0$, which trivially implies the postcondition $0 \leq v$. But the same cannot be said about the first conjunct with the postcondition $x \leq m$. Indeed, after some more decompositions by logical equivalences, it will turn out to be false if the initial velocity of the car exceeds its braking capabilities compared to the distance to the obstacle m.

The big conceptual advantage of working with logic for cyber-physical systems is that one can start with an unambiguous question phrased in dL and then subsequently transform it with logic such that every step along the way is easy and clearly correct and the final outcome is easier than the original question. It is, indeed, in many ways due to the use of logic that the LFCPS course is successful in simplifying the otherwise overwhelming challenges of cyber-physical systems by reducing them to simpler pieces [13].

Active Learning Quizzes. The LFCPS course features active learning quizzes for every chapter and lecture of the accompanying textbook [15]. Learning by doing is a crucial element of understanding material. The purpose of the course quizzes is to support the student's learning by giving them an opportunity to practice and get feedback on how well they have achieved a selection of some of the learning goals of the LFCPS course. By observing which ones the student is unsure about, the quizzes can help identify which material they should review again. Since students ultimately need a solid understanding of all aspects of CPS, this helps stay up to speed.

The most profound impact of student learning stems from the ways of thinking that is internalized so deeply that the student can produce them on the fly without having to look anything up. Concepts that become part of their thinking will enable students to autonomously detect situations where they apply, instead of needing to rely on others to tell them which concept to apply in order to solve which problem.

While quizzes feature carefully paced introductory questions, they are also designed to challenge a student's understanding. This gives them an opportunity to think through some of the more subtle aspects of CPS at their own pace before they face similar challenges in application contexts where challenges may become overwhelming. By solving a sequence of such separate challenges, students become better at understanding nuances and internalize the way of thinking that is required to solve them. A few of the quiz questions give students an opportunity to synthesize multiple individual concepts to solve a small joint challenge. These questions exercise synthetic knowledge and enable students to form conceptual bridges between individual skills to identify what they need where.

For example, some of the quiz questions ask students to check their thinking on certain simple subskills, which are useful to acquire early to avoid confusions. Other quiz questions may make them wonder how long differential equations evolve and what exactly a safety property of a hybrid system means. These are fundamental questions about CPS models that they can answer using their semantics. Yet other quiz questions ask students to put all their acquired skills together to design simple CPS controllers or criticize their designs before facing the challenges of real applications. Discovering a problem in one's thinking in the small context of a quiz question is a great learning experience and prevents students from the major downstream effects of carrying a conceptual misunderstanding forward into later parts of the course. Quiz questions make students confront the blank page syndrome in the small, where they are asked to creatively come up with answers to small questions on their own. This experience is not easy but prepares students for when they face bigger challenges where they will creatively come up with answers to bigger questions.

Except for the summary and wrapper questions with free text answers that are included for the purpose of reflection and feedback, the LFCPS active learning quizzes are fully autograded by the theorem prover KeYmaera X [4] that proves correctness of the student's answer automatically and giving some question-dependent feedback when this fails. It only happened twice that a student provided a correct answer on the active learning quiz that KeYmaera X was unable to prove. By having even thought about this, the student arguably learned more than what could ever be reflected in the 2 missed points out of 500 quiz points.

Example Quiz Questions. The easiest quizzes to design are multiple-choice quizzes. But those only teach students the passive skill of recognizing the right answer rather than the active skill of creatively producing the right answer. This is essentially the educational counterpart to the computational P-NP problem [2]. Checking correct answers is easier than producing correct answers. One cannot learn integration by multiple-choice, because the mere process of differentiating the given answers masks the intended ability to learn how to integrate functions. To give the reader a feeling how the LFCPS active learning quizzes feature genuinely *active* learning, here is a small sampling of typical question types.

Example 4 (Quiz: Program shapes). Objective (*programming languages for CPS, semantics, models, operational effects*): It is crucial to obtain an intuitive reading of the respective transitions in a hybrid program. This question gives you an opportunity to practice the mapping between a transition structure and the hybrid programs they correspond to.

What hybrid program fits to the following transition structure?

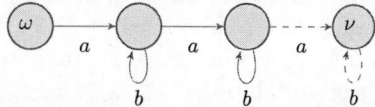

Answer: {{a;{b;}*}*
Similar to an ellipsis, the dashed part of the transition systems indicates the transitions as shown that already happened before may happen again and again. Every a can be followed by an arbitrary number of b, as briefly indicated by the self loop with *b*, and the whole pattern can repeat arbitrarily often, as indicated by the dashed parts. Come to think of it, if the b transition goes back with a self-loop literally to the exact same state, then it cannot have had a huge effect or any at all. But, instead, the diagram illustrates the structure of the transition so the self-loops indicate that b can happen repeatedly at the indicated nodes, so {a;b;}* would be incorrect, because it does not capture the fact that there can be an arbitrary number of b (even 0) between two consecutive a occurrences. The answer a;{b;}*;a;{b;}*;{a;{b;}*}* would fit to the above transition structure as well, but is unnecessarily complicated, and, thus, not an insightful answer.

For each of the following transition structures, find a hybrid program that can mimic its decisions. Give the simplest (shortest) hybrid program that can mimic all actions in the transition structure.

$$\omega \; \underset{a}{\to} \; \bigcirc \; \underset{b}{\to} \; \omega_1 \; \underset{a}{\to} \; \bigcirc \; \underset{b}{\to} \; \omega_2 \; \underset{a}{\to} \; \bigcirc \; \underset{b}{\to} \; \nu$$

The quiz question type in Example 4 inverts definitions, giving students an opportunity to understand constructions on a deeper level by asking what input to a semantics definition would have a shape that fits to the expected output. While the forward application of definitions is a crucial skill, the backward or inverse application requires reflection and practices deeper understandings of the inner working principles. Without techniques for computing integrals, finding a function that has a given derivative requires a much deeper understanding about the process of differentiation.

Example 5 (Quiz: True formulas). Objective (*model semantics, preconditions, rigorous specification*): If a formula is not valid, it is important to identify when exactly it is true. This helps identify missing preconditions to make it valid, and read off consequences when a formula is available as an assumption. Of course, knowing when exactly a formula is true is also crucial when they are used as evolution domain constraints or tests, which is why those are usually quantifier-free FOL formulas.

Question: When is dL formula $[x := x + 1] x > 5$ true?

Answer: x>4

Indeed, the first formula $[x := x + 1] x > 5$ and the second formula $x > 4$ are equivalent, i.e., they are true in exactly the same states, because the value of x after the assignment $x := x + 1$ is one larger than it was before. So $x > 5$ after $x := x + 1$ iff $x > 4$ initially. But the latter formula $x > 4$ is easier to understand than its equivalent $[x := x + 1] x > 5$, because it merely involves arithmetic evaluation, rather than also effectful programs.

For each of the following dL formulas identify the **exact** set of all states in which it is true and characterize this set by a quantifier-free formula of real arithmetic (of the same free variables).

$$[x' = v, v' = a] x \leq m$$

The quiz question type in Example 5 asks students to creatively come up with the answer when they learn something about continuous dynamics to characterize when exactly a car always stays before position m. The answer is not easy but gives students a chance to practice and check their physical intuition as a logical formula. Answering it helps ultimately fix and prove Example 1 in a subsequent synthetic skill question in a later quiz. While answering this quiz question, students are led down a corridor of exploration of increasingly refined understandings of the safety of continuous dynamics.

Example 6 (Quiz: Axiom usage). Objective (*rigorous reasoning about CPS*): As one important part of rigorous reasoning about CPS, you will practice the correct application of axioms to differential dynamic logic problems. While the KeYmaera X prover correctly applies axioms for you, it is still helpful if you practice this yourself to get a better intuition for how it works and predict the outcome of a proof step before trying it. That will make you more time-efficient in your reasoning. It will also inform you how to transform parts of a proof to make useful axioms applicable later. If you properly understand reasoning principles, you are also better able to identify and check clever problem decompositions.

Question: What is the result of using axiom [;] on $[ctrl; plant] x > y$?

Answer: `[ctrl;][plant;]x>y`

Question: What is the result of using axiom [:=] on $[ctrl; plant] x > y$?

Answer: `n/a`

Because the axiom [:=] expects an assignment instead of a ; as the top-level operator in the box modality, so is not applicable to the given formula, which is not of the form expected by the left-hand side of axiom [:=].

For each of the following dL formulas, give the result of using the indicated dL axiom (in its usual left to right decomposition direction) **once** to the whole formula **at its top-level position**, i.e., for the top-level operator and not deep down in the middle of some subformula. Respond n/a if the axiom is not applicable to the given formula.

Axiom [;] on [sense; ctrl; plant] $x > y$

The quiz question type in Example 6 teaches students the precision of reasoning by asking them to predict a proof step of a theorem prover. One might think that this skill is irrelevant, because a theorem prover such as KeYmaera X [4] implementing differential dynamic logic will be perfectly capable of applying proof principles correctly. But it is also helpful to develop an intuition for what will happen when, be able to predict the outcome, appreciate the nuances of reasoning overcome by a theorem prover along the way, and, especially, be able to do and check manual paper-based proofs as well. Along the way of answering the above quiz question, students will discover on their own the significance of precision in the syntactic representation of programs.

Example 7 (Quiz: Axiom development). Objective (*operational CPS effects,* dL *as a verification language*): The axioms of differential dynamic logic are complete, so you do not need any more for its operators. But whenever you add new syntax to the language, then you give that operator a semantics, and also need to add new axioms for reasoning about the new syntactic features. These questions give you an opportunity to practice the extension of syntax, semantics, and axiomatics that fit together in harmony and properly decompose hybrid programs into logic. Recall that it is imperative that only sound axioms be adopted. Also remember that solid axioms for a program statement reduce the new syntactic program operator to simpler logical formulas about subformulas and subprograms, because that makes it possible to understand the new program operator solely in terms of easier logic, not in terms of different or more complicated programs. This question gives you the opportunity to practice the development of new axioms for hybrid programs.

The if-then-else statement $\text{if}(Q)\,\alpha\,\text{else}\,\beta$ runs HP α if formula Q is initially true and runs HP β otherwise. Its semantics is $\ll \ldots \text{elided} \cdots \gg$.

Develop an axiom for $[\text{if}(Q)\,a\,\text{else}\,b]P$ that decomposes the effect of the if-then-else statement in logic with simpler logical connectives.

After the axiom reflection question Example 6, and subsequent questions that ask students to check the correctness of conjectured axioms, the quiz question type in Example 7 asks students to create rather than use axioms for reasoning about CPS, which enables them to develop higher metacritical analytic skills.

Example 8 (Quiz: Loop invariants). Objective (*identifying and expressing invariants*): The most important ingredient of a CPS is its invariant, because an invariant tells you what you always know about your system, no matter how long it operates. This question allows you to practice the important but challenging task of identifying loop invariants for hybrid systems.

Identify a loop invariant J proving the following dL formulas (after using rule $\to$R) with exactly the following version of the loop invariant proof rule:

$$\text{loop} \ \frac{\Gamma \vdash J, \Delta \quad J \vdash [\alpha]J \quad J \vdash P}{\Gamma \vdash [\alpha^*]P, \Delta}$$

Write n/a when no loop invariant exists that proves the given dL formula.
$$x \geq 1 \wedge v > 0 \wedge A > 0 \to [((a := 0 \cup a := A); \{x' = v, v' = a\})^*] x \geq 0$$

The quiz question type in Example 8 implicitly follows up on the skills developed in Example 5 and gives students the opportunity to figure out why a control loop works correctly. They are not kept guessing whether their answer is correct, because the KeYmaera X theorem prover underlying the active learning quizzes tells them right away when the proof did not succeed. Other questions practice reflection and understanding by asking for the shortest possible invariant of a controller or differential equation, which helps students understand which arguments are necessary compared to which arguments are true but useless. Subsequent quiz questions practice synthetic skills for system design and safety justifications while helping students understand the beneficial role that incremental system designs play in reducing verification and comprehension challenges.

CPS V&V Grand Prix. In many but not all years, the LFCPS course also features practical verification labs culminating in self-defined course projects. The students design, analyze, and prove correct a sequence of designs for robot models mastering increasingly difficult challenges. In each lab, the students design a controller for a single robot that can interact with an unknown environment. The students also design an appropriate model for the continuous behavior that their controlled robot would exhibit given the discrete control inputs. They need to decide on an appropriate model for the robot's environment, including using nondeterminism to capture unknown behaviors in the environment. And finally, the students formalize a safety property as a logical formula and prove that their controller never violates it. The labs are all related and build on each other, with the ultimate goal that the students design and prove safety for a robot that can avoid moving obstacles.

Before students submit the final robot model (called *Veribot*), they submit a *Betabot*, which is a beta-version of the robot controller that they conjecture to be safe and submit for feedback. Unlike the final robot submission (the Veribot), the Betabot does not yet need to be verified, but should provide best-thought-out conjecture in order to give students a head-start on the Veribot. This teaches students by experience that most CPS designs are more challenging than it appears at first glance. Students learn to appreciate the value of formal verification by seeing first hand the quality difference between their Betabots and their ultimate verified Veribots. Feedback on the Betabots also prevents students from wasting effort on models that have fundamental flaws.

The course culminates in a self-defined final course project, with which the students compete in the CPS V&V Grand Prix course competition, presenting to a panel of about 12 experts in CPS who give them feedback from an industry perspective from organizations such as Siemens, Google, Bosch, Aptiv, Galois, Argo AI, Near Earth Autonomy, MathWorks, Toyota, NASA, Intel, GM who also award prizes. The mix of substantial conceptual challenges and self-defined course projects with written reports and slide presentations in the competition for the industry judges gives students a chance to shine and the instructor and judges a chance to get to know them fairly well. Because of the time pressure and random effects of presentations at the competition, as well as mix of undergraduates, master's students, and PhD students competing, the final rank in the Grand Prix is no direct indicator of the overall skill of a student. In fact, the more detailed study of their course projects by my TAs and me frequently shines a complementary and more thorough light on their innovation and technical quality that they did not communicate perfectly to the judges. But this is overall a much-appreciated factor of the LFCPS course that it provides so many different ways to shine and become a verification rock-star.

Evaluation. While the course evaluations are essentially perfect (with the exception of understandable concerns about the high workload), a much more useful indicator of success is the fact that so many students take this course, despite the fact that it is much easier to earn a good grade in other courses, and even though, when run with all its components, the course has a very intensive workload. The author is since experimenting with ways of reducing the workload without over-proportionally reducing the learning outcomes.

Anecdotal evidence since the introduction of active learning quizzes into the LFCPS course led to students who are significantly advanced compared to their peers from prior years. This experiment is imperfect, however, because active learning quizzes were introduced at the start of the covid-19 pandemic, where the course became fully remote, so that more than one aspect of the course was changed at the same time. Statistical analysis also indicates that the active learning quiz scores are the strongest predictor for the overall course grade.

4   Course: Programming Language Semantics

The Programming Language Semantics course that the author taught at Carnegie Mellon University was a logical redesign benefitting significantly from previous course designs by Steve Brookes and John Reynolds [17]. The use of logic in studying programming language semantics is universal wisdom, most obviously in the case of axiomatic semantics for programming languages, which establishes logical axioms that characterize the truth of statements about programs. Other parts of programming language semantics courses deal with several variations of denotational and operational semantics. The relation between those different flavors of semantics is most exciting and valuable. The equivalence of denotational and operational semantics, which helps combine the advantages of both, and the soundness and relative completeness relations of denotational or operational and axiomatic semantics, teach valuable lessons about how to justify that different perspectives on a programming language are in sync. But different designs of programming language semantics courses differ in how pervasively logic is used. Tobias Nipkow, for example, propagates the pervasive use of the theorem provers Isabelle/HOL to study programming language semantics [5,6]. The author's Programming Language Semantics (PLS) course does not use formal theorem provers, which reduces the learning curve of proof tools, but still consistently benefits from the use of dynamic logic and uniform substitution [14].

Partial Semantics. One obvious difference impacted by the use of logic is in the development of program states. For PLS courses that are developed from the perspective of programming languages, it is natural to define states as partial functions $\omega : \mathcal{V} \dashrightarrow \mathcal{D}$ from variables in $\mathcal{V}$ to values in $\mathcal{D}$ and then define the meaning of expressions and programs and assertions about programs equally partially while distinguishing cases on whether or not a state is defined on all variables that a program reads or that a formula depends on. For instance, the semantics $\omega[\![x+1]\!]$ of a program expression $x+1$ in the state ω is only defined if the state ω actually gives a value to the variable x that the expression $x+1$ mentions. Otherwise, $\omega[\![x+1]\!]$ remains undefined. This detail is useful to teach about delicacy and precision, and prepares for other reasons of undefinedness in program execution such as null-pointer dereferences or out of bounds array accesses, but it does cause a fair amount of technical hurdles and complications at every step along the way if every connection of syntax and semantics needs to be guarded by an assumption that the states are defined everywhere where they need to be to make sense of the syntax.

Impartial Semantics. For PLS courses developed from the perspective of logic, it is more natural to define states as total functions $\omega : \mathcal{V} \to \mathcal{D}$ from variables to values, which makes it easier to evaluate terms and formulas and programs as a function of the values of the variables they read. The semantics $\omega[\![x+1]\!]$ is always defined and equas $\omega(x)+1$, since the total function ω has some value $\omega(x)$ for every variable $x \in \mathcal{V}$. The resulting substantial conceptual simplicity is a benefit even if states "waste information" in the sense that variables have values even if they are never needed. With the establishment of an easily proved coincidence lemma, saying that the value of formulas and the effect of programs only depends on the values of their free variables $FV(\cdot)$, this difference between both approaches vanishes, but the conceptual simplicity and absence of technicalities of undefinedness remains a benefit of the approach coming from logic.

Lemma 1 (Coincidence lemma [14]). *If $\omega = \nu$ on $FV(\theta)$, then $\omega[\![\theta]\!] = \nu[\![\theta]\!]$.*

After establishing this coincidence lemma it becomes clear retroactively that only the values of some part of the state affect the meaning of terms θ (and similarly for other expressions). But undefinedness or the need to check for the semantic compatibility of states and expressions was never a concern.

Syntactic Transformation via Special-purpose Semantics. Even more pronounced is the simplicity afforded by logic for program transformations and program contexts. A natural operation on programs is to replace one part with something else. Obviously, some program transformations change the program behavior while others do not. Proving when program transformations do not affect the program behavior is surprisingly difficult without the use of logic even in simple cases. To illustrate, a transformation turning a part e of a program into k conventionally needs the definition of a program context as a program $\alpha(\_)$ with a hole $\_$ that is once filled with the syntactic part e to form $\alpha(e)$ and that is once filled with the syntactic part k to form $\alpha(k)$ giving the following rewrite:

$$\alpha(e) \rightsquigarrow \alpha(k)$$

Besides defining the syntax of programs, the syntax of programs with holes, the syntactic operation of the program transformation, a correctness argument also needs to define not just the semantics $[\![\alpha]\!]_{\text{prg}}$ of programs α but also the semantics $[\![\alpha(\_)]\!]_{\text{cxt}}$ of programs $\alpha(\_)$ with holes, and the semantics $[\![e]\!]_{\text{part}}$ of the affected parts e. And one then has to prove that the semantics of the program obtained after the syntactic transformation of plugging in e into the program $\alpha(\_)$ with hole $\_$ to obtain the program $\alpha(e)$ is equivalent to the semantics $[\![\alpha(\_)]\!]_{\text{cxt}}$ of the program with hole $\alpha(\_)$ applied to the semantics $[\![e]\!]_{\text{part}}$ of the part e:

$$[\![\alpha(e)]\!]_{\text{prg}} = [\![\alpha(\_)]\!]_{\text{cxt}}([\![e]\!]_{\text{part}}) \tag{1}$$

Of course, inductively proving this particular compatibility result (1) teaches one to scrutinize and relate several different but closely related versions of semantics definitions with inductions on the syntax and corresponding decompositions on the semantics. But the proof of (1) and similar results is fairly technical and all the required definitions repetitive even if the semantic domain shifts a little from program behavior to functions from hole-filling behavior to program behavior. Likewise one will have to prove that the semantics of a program context α that does not, in fact, even have a hole is actually equivalent whether taken as a program or as context semantics:

$$[\![\alpha]\!]_{\text{prg}} = [\![\alpha]\!]_{\text{cxt}}(S) \quad \text{for any possible semantics } S \text{ that a part } e \text{ may have}$$

Syntactic Transformation via Logic. Taking logic seriously leads to another angle on PLS courses, where the use of dynamic logic and uniform substitution significantly simplify technical challenges. Uniform substitution was originally defined for first-order logic by Church [1, §35,40] for substituting function symbols with terms and substituting predicate symbols with formulas, uniformly everywhere, while respecting that free variables cannot be bound during the substitution [14]. Corresponding generalizations lift uniform substitution to dynamic logic with programs, where, in addition, program symbols can be substituted with programs [14]. The uniform substitution proof rule US states that if a formula φ has a proof then all its uniform substitution instances $\sigma(\varphi)$ have a proof.

Theorem 1 (Soundness of Uniform Substitution [14]). *The proof rule US is sound (where US is only applicable if the substitution result $\sigma(\varphi)$ is defined).*

$$\text{US } \frac{\varphi}{\sigma(\varphi)}$$

Proving soundness of the uniform substitution proof rule US takes some effort as well but only needs to be established once and for all as the *only form of syntactic transformation*. A similar proof principle uses the same uniform substitution σ simultaneously on all premises and the conclusion of a (locally sound) inference.

Theorem 2 (Soundness of Uniform Substitution of Rules [14]). *All uniform substitution instances (whose substitutions introduce no free variables) of locally sound inferences are locally sound:*

$$\frac{\phi_1 \quad \cdots \quad \phi_n}{\psi} \text{ locally sound} \quad \text{implies} \quad \frac{\sigma(\phi_1) \quad \cdots \quad \sigma(\phi_n)}{\sigma(\psi)} \text{ locally sound}$$

Correctness of the syntactic transformation of term e for an equal term k is then a uniform substitution instance of the following obvious axiomatic proof rule:

$$\text{CQ } \frac{f = g}{p(f) \leftrightarrow p(g)}$$

The congruence rule CQ states that if (nullary function symbol) f is equal to g then for any (unary) predicate p the formula $p(f)$ is equivalent to $p(g)$, which is evidently correct by the principle of substitution of equals for equals. Making a concrete inference with rule CQ simply amounts to using the uniform substitution principle from Theorem 2 to substitute the concrete term e for the function symbol f and the concrete term k for the function symbol g and a concrete property of a concrete program for the predicate symbol p in which, by way of the principle that replacements of predicate symbols will still have their arguments in the same places, implicitly defines a context without the need for separate definitions, semantics and constructions. Thanks to uniform substitution, which modularizes all the challenges of how syntactic transformations preserve the semantics, independently of the particular use case, congruence reasoning and contextual replacements reduce to the self-evident rule CQ.

The same principle of substituting equals for equals immediately transfers to the situation when replacing formulas for equivalents or when replacing subprograms without the need to define a new semantics and new syntactic replacement principles and their semantic compatibility from scratch. For example, congruence on terms reduces to uniform substitution uses via Theorem 2 of the evident axiomatic proof rule CT:

$$\text{CT } \frac{f = g}{c(f) = c(g)}$$

No new syntactic category of terms with holes or new semantics or new replacement mechanisms or new compatibility proofs are needed. Everything simply follows from uniform substitution.

Logical Compiler Optimizations. Likewise, program transformations such as common subexpression elimination are but uniform substitution instances of the backwards direction of the assignment axiom:

$$[:=]\ [x := e]p(x) \leftrightarrow p(e)$$

The idea is merely that a common subexpression e identified in a formula $p(e)$ (or a program contained therein for the purpose of showing equivalence of the transformation) that may occur multiple times in different places is pulled out and stored into a common variable x that is assigned to after computing the value of the term e only once.

Example 9 (Common subexpression elimination via logic). The common subexpression $a^2 + b$ can be pulled out of the following program

$$[\text{while}(y^2 < a^2 + b)\ \{z := z + y^2 * (a^2 + b); y := y + 2 * 3\}]P$$

to obtain the, by $[:=]$ equivalent

$$[x := a^2 + b; \text{while}(y^2 < x)\ \{z := z + y^2 * x; y := y + 2 * 3\}]P$$

But common subexpression elimination cannot pull out any of the y^2 occurrences, because y is changing afterwards in the loop body which will be before the (textually earlier) next use of y in the next round of the loop. This difference of applicability of common subexpression elimination, which is crucial for correctness, is easily spotted by uniform substitution via Theorem 1 applied to axiom $[:=]$. Of course, another intuitive giveaway is that, if all parts of a loop condition were common subexpressions, then the loop condition never changes its truth value, so the loop either never runs or runs forever.

Ordinarily, special-case analyses would have to be designed and their correctness proven separately for syntactic transformations such as common subexpression elimination in programming language compilation. The consequent use of logic reduces this to mere uniform substitution.

Example 10 (Copy propagation via logic). Copy propagation is another compiler optimization that, of course, needs soundness-critical applicability checks. Propagating the value of a variable to later occurrences of that variable is again just uniform substitution via Theorem 1 for the $[:=]$ axiom. This, for example, rephrases the last program of Example 9 equivalently to

$$[x := a^2 + b; \text{while}(y^2 < x)\ \{z := z + y^2 * (a^2 + b); y := y + 2 * 3\}]P$$

or to

$$[x := a^2 + b; \text{while}(y^2 < a^2 + b)\ \{z := z + y^2 * (a^2 + b); y := y + 2 * 3\}]P$$

Again, uniform substitution directly tells apart this correct use of copy propagation from an incorrect attempt to propagate the value $z + y^2 * x$ for any occurrence of z (such as after the program or in itself), because that would already have a different value due to the loop.

Example 11 (Constant folding via logic). The constant folding transformation is merely a use of uniform substitution for the congruence rule CQ, e.g., to reduce that the multiplication $2 * 3$ is 6 in a program, because $2 * 3 = 6$ is true. This reasoning rephrases the last program of Example 9 equivalently to

$$[x := a^2 + b; \text{while}(y^2 < x) \{z := z + y^2 * x; y := y + 6\}]P$$

If, for some reason, one would like to commute $a^2 + b$ to $b + a^2$ (maybe in order to enable more common subexpression eliminations in other parts of the program), then another use of uniform substitution for congruence rule CQ uses the equation $a^2 + b = b + a^2$ to transform the above program equivalently:

$$[x := b + a^2; \text{while}(y^2 < x) \{z := z + y^2 * x; y := y + 6\}]P$$

With slight generalizations of uniform substitution to program relations such as refinement $\alpha \leq \beta$ and equivalence $\alpha = \beta$ [16], one can state a congruence rule for making use of program equivalences in any context $C(\_)$

$$\text{CP} \quad \frac{\alpha = \beta}{C(\alpha) \leftrightarrow C(\beta)}$$

Example 12 (Loop unwinding via logic). Unwinding one round of a loop to run before the loop is easily done via uniform substitution via Theorem 2 on congruence rule CP. Based on the equivalence

$$\text{while}(Q)\,\alpha = \text{if}(Q)\,\{\alpha; \text{while}(Q)\,\alpha\}$$

the last program of Example 11 is transformed via CP to its equivalent:

$$[x := b + a^2; \text{if}(y^2 < x)\,\{z := z + y^2 * x; y := y + 6;$$
$$\text{while}(y^2 < x)\,\{z := z + y^2 * x; y := y + 6\}\}]P$$

If interesting optimizations now were to become possible in the pulled out first iteration, subsequent combinations of logical transformations can continue. Here, this might apply, e.g., if the initial value of y is known to satisfy $y^2 < b + a^2$ such that the if branching is unnecessary.

5   Conclusion and Outlook

Overall, symbolic logic plays and should play a significant role in scientific education. Logic both leads to significant simplifications of otherwise challenging concepts and logic leads to identifying the inherent core essence of ideas otherwise lost among lots of peripheral aspects. As the experience relayed in this paper demonstrates, there is a wide variety of courses that benefit from the inclusion of ideas from logic ranging from introductory courses over systems courses to logic and programming language courses. Logic helps on a scale ranging from tool-free informal logic reasoning supported by practice with mere dynamic checking of program-expressed contracts all the way to full feature theorem provers. Active learning quizzes with theorem provers fully blackboxed in its autograder have an outside impact on the understanding of students that seems well worth the nonnegligible time investment. The development of pedagogically well-paced and insightful questions aligned with the learning goals of the course is, however, a massive undertaking that can only be amortized by reusing the quiz questions and autograding infrastructure over the years.

References

1. Church, A.: Introduction to Mathematical Logic. Princeton University Press, Princeton (1956)
2. Cook, S.A.: The complexity of theorem-proving procedures. In: Harrison, M.A., Banerji, R.B., Ullman, J.D. (eds.) STOC, pp. 151–158. ACM, New York (1971). https://doi.org/10.1145/800157.805047
3. Egerstedt, M., Gupta, R., Jensen, J.C., Lee, E.A. (eds.): CPS-Ed (2013)
4. Fulton, N., Mitsch, S., Quesel, J.-D., Völp, M., Platzer, A.: KeYmaera X: an axiomatic tactical theorem prover for hybrid systems. In: Felty, A.P., Middeldorp, A. (eds.) CADE 2015. LNCS (LNAI), vol. 9195, pp. 527–538. Springer, Cham (2015). https://doi.org/10.1007/978-3-319-21401-6_36
5. Nipkow, T.: Teaching semantics with a proof assistant: no more LSD trip proofs. In: Kuncak, V., Rybalchenko, A. (eds.) VMCAI 2012. LNCS, vol. 7148, pp. 24–38. Springer, Heidelberg (2012). https://doi.org/10.1007/978-3-642-27940-9_3
6. Nipkow, T., Klein, G.: Concrete Semantics. Springer, Cham (2014). https://doi.org/10.1007/978-3-319-10542-0
7. Perlis, A.J.: Special feature: epigrams on programming. ACM Sigplan Notices **17**(9), 7–13 (1982). https://doi.org/10.1145/947955.1083808
8. Platzer, A.: Differential dynamic logic for hybrid systems. J. Autom. Reas. **41**(2), 143–189 (2008). https://doi.org/10.1007/s10817-008-9103-8
9. Platzer, A.: Differential Dynamic Logics: automated Theorem Proving for Hybrid Systems. Ph.D. thesis, Department of Computing Science, University of Oldenburg (2008)
10. Platzer, A.: Logical Analysis of Hybrid Systems: Proving Theorems for Complex Dynamics. Springer, Heidelberg (2010). https://doi.org/10.1007/978-3-642-14509-4
11. Platzer, A.: Logics of dynamical systems. In: LICS, pp. 13–24. IEEE, Los Alamitos (2012). https://doi.org/10.1109/LICS.2012.13
12. Platzer, A.: Teaching CPS foundations with contracts. In: CPS-Ed, pp. 7–10 (2013)
13. Platzer, A.: Logic & proofs for cyber-physical systems. In: Olivetti, N., Tiwari, A. (eds.) IJCAR 2016. LNCS (LNAI), vol. 9706, pp. 15–21. Springer, Cham (2016). https://doi.org/10.1007/978-3-319-40229-1_3
14. Platzer, A.: A complete uniform substitution calculus for differential dynamic logic. J. Autom. Reas. **59**(2), 219–265 (2017). https://doi.org/10.1007/s10817-016-9385-1
15. Platzer, A.: Logical Foundations of Cyber-Physical Systems. Springer, Cham (2018). https://doi.org/10.1007/978-3-319-63588-0

16. Prebet, E., Platzer, A.: Uniform substitution for differential refinement logic. In: Benzmüller, C., Heule, M., Schmidt, R. (eds.) IJCAR. LNCS, vol. 14740, pp. 196–215. Springer, Cham (2024). https://doi.org/10.1007/978-3-031-63501-4_11
17. Reynolds, J.C.: Theories of Programming Languages. Cambridge University Press (1998). https://doi.org/10.1017/CBO9780511626364
18. Wigner, E.P.: The unreasonable effectiveness of mathematics in the natural sciences. Commun. Pure Appl. Math. **13**(1), 1–14 (1960). https://doi.org/10.1002/cpa.3160130102
19. Wing, J.M.: Computational thinking. Commun. ACM **49**(3), 33–35 (2006). https://doi.org/10.1145/1118178.1118215

Regular Papers

Introducing GitHub Classroom into a Formal Methods Module

Soaibuzzaman[(✉)] and Jan Oliver Ringert[(✉)]

Bauhaus-University Weimar, Weimar, Germany
`jan.ringert@uni-weimar.de`

Abstract. We have developed an MSc-level module on Formal Methods for Software Engineering with exercises on applying SAT solvers, SMT solvers, Alloy, and nuXmv. In the first iteration of the module, assignments were submitted as documents and archive files. Here, we report on our experience of moving the exercises to GitHub Classroom and automating the feedback process through test cases. The main challenges we encountered were related to supporting free-response tasks and designing test cases that allow for multiple solutions, provide incremental feedback, and do not encode a solution. We present our setup of exercise repositories, test cases, and feedback report generation. We detail our approach in addressing the challenges of migrating from worksheets to GitHub Classroom and report on survey-based student feedback.

Keywords: teaching · formal methods · GitHub classroom · feedback

1 Introduction

We have developed an MSc-level module on Formal Methods for Software Engineering with exercises on applying SAT solvers [4,20], SMT solvers [3,23], Alloy [15], and nuXmv [5]. As well recognized by others, practical examples and exercises are important when teaching Formal Methods [6–8,19,26]. Our module contains five chapters, where each technical chapter (2–5) is supported by a specification and an implementation exercise (see Sect. 2). In specification exercises, students learn to use specification languages for smaller examples. In implementation exercises, students automate the programmatic translation of domain problems to solvers.

In the first iteration of the module, exercises were submitted as documents and archive files. Here, we report our experience moving the exercises to GitHub Classroom (GHC). Our main goals were to improve the student learning experience [10,28,32] by using autograding and providing immediate feedback [22] and to shorten grading times and teacher resources.

We present our setups of exercise repositories, test cases, and feedback report generation in Sect. 4.2. Our main challenges were supporting free-response tasks and designing test cases that allow for multiple solutions, provide incremental feedback, and do not encode a solution (see Sect. 4.3). We detail our approach

E. Sekerinski and L. Ribeiro (Eds.): FMTea 2024, LNCS 14939, pp. 25–42, 2024.
https://doi.org/10.1007/978-3-031-71379-8_2

to addressing the challenges of migrating from worksheets to GHC. We report on survey-based student feedback collected after each exercise in the year of the migration to GHC in Sect. 5.

2   Module Overview

We have developed the module Formal Methods for Software Engineering as an MSc-level module of 6 ECTS (roughly 180 h of student workload in one semester). The module is offered to students from three MSc degree programs at Bauhaus-University Weimar: Computer Science for Digital Media, Human-Computer Interaction, and Digital Engineering (most engineering and computer science backgrounds with mandatory programming experience and a total of 36 ECTS in computer science subjects). Our module has no formal prerequisites, but we expect that students have a BSc-level background in mathematics and basic programming knowledge. Additionally, Software Engineering and Object-Oriented Programming is recommended for Digital Engineering students from non-computer-science backgrounds.

An outline of the module is shown in Fig. 1 with five main chapters spread over 15 weeks. The first chapter *Declarative thinking* motivates the use of specifications for SE. Each following chapter introduces a logic, a language, a tool, and applications, e.g., the second chapter *SAT solving* introduces propositional logic, the syntax of limboole [4], and limboole; the final chapter *Model-checking* introduces Linear Temporal Logic [25], the SMV language, and the nuXmv tool [5].

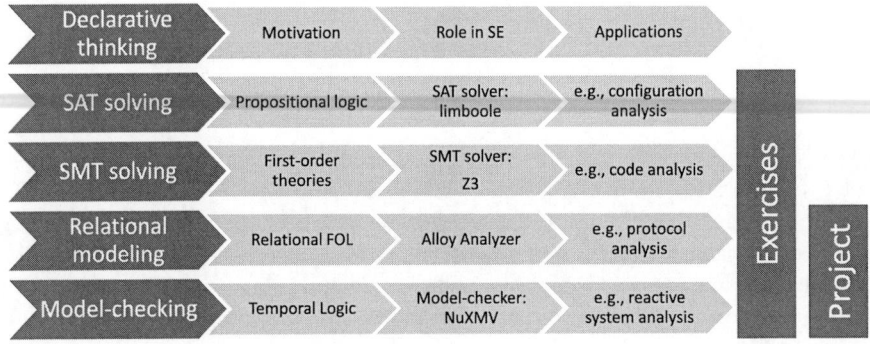

Fig. 1. Outline of Formal Methods for Software Engineering divided into five chapters

The module's content and intended learning outcomes focus on selecting and applying specification formalisms and tools to address software engineering challenges rather than on the tools' underlying algorithms and decision procedures. The module is assessed 100% based on a project. Passing all seven exercises is mandatory before submitting a project. Two extensions or resubmissions (out of seven) are granted for failed exercises. Students work in pairs or alone.

2.1   Exercise Structure

Chapters 2–5 from Fig. 1 were each supported by a specification exercise (one week to complete) and an implementation exercise (two weeks to complete).

Specification exercises require manually writing specifications and encoding problems, e.g., solving a math puzzle using SMT as shown in Fig. 2 (Ex. 3, Task 2) or determining semantic equivalence of two LTL formulas using the nuXmv [5] model-checker. The goal of these exercises was to learn the features of the new language and to be able to express puzzles and problems in each formalism.

Implementation exercises require the automation of the encoding and the solving tasks as a program, e.g., finding dead features in a feature model [21] (SAT solving) or configuring a PC system from components with different prices, budget limitations, and configuration constraints (SMT solving).

We chose Java as a common language known to most students. The generation of input for the solver was based on String manipulations for SAT solving and APIs for SMT (JavaSMT [3]) and Alloy [15]. We did not provide an implementation exercise for the *Model-checking* chapter, as the students focused on their projects. The goal of these exercises was to learn the generalization of problems and the automation of systematically encoding and solving them.

We provide an overview of exercises and present some excerpts of the worksheets in Fig. 2. All exercise materials are available for inspection, modification, and reuse (Apache 2.0 license) from https://github.com/fm4se/exercises/:

1. SAT *spec*: formulas, checking conclusions, verifying Role-Based-Access
2. SAT *impl*: Feature Model analysis [21], dead features, product preservation
3. SMT *spec*: Agatha puzzle [24, P.55], math puzzle, PC configuration
4. SMT *impl*: PC configuration from CSV-files, budget and purpose
5. Alloy *spec*: domain model, Agatha puzzle [24, P.55], Trash can [19]
6. Alloy *impl*: Analysis of Alloy modules: dead signatures, minimal scopes
7. nuXmv *spec*: LTL equivalence, counterexamples, chess knight moves

3   Related Work

3.1   GH in Classroom

Numerous studies have evaluated the use of GitHub and GitHub Classroom for teaching programming courses. Hsing and Gennarelli [14] conducted a study exploring the advantages of incorporating GitHub into the classroom. Their research revealed that students who received feedback through GitHub achieved better learning outcomes and developed the necessary skills for collaborative work. Their survey included 7,530 students and 300 educators. Furthermore, a group of instructors from the University of Auckland [32] shared their experience using GitHub Classroom. They implemented Git and related systems in various courses with varying class sizes, experience levels, and contexts. Their findings indicated that while introducing GitHub to the classroom increased the

---

Ex. 3, Task 2

Encode the Puzzle:

(🪙 * ♟️) + ⚙️ = 110

(⚙️ - ♟️) + ❤️ = 17

(♟️ * 🪙) - ♟️ = 90

(⚙️ - 🪙) - ❤️ = ?

Note: use the same names as in the template to encode the puzzle.

Ex. 3, Task 3

Create a reusable encoding of the selection of PC components (see below table) and the calculated price of a PC.

- The selection must satisfy the listed constraints.
- Each component may only be selected once, i.e., you cannot include two CPUs.

The encoding should be reusable in the following sense:

- A user states a purpose and budget and the encoding produces models that represent PC configurations matching the requirement, if possible.
- Different purposes add further constraints listed below.

Use the following template to start with. It already contains the encoding of the variables budget and purpose. Keep these names.
Constraints:

- Each computer needs basic components: CPU, Motherboard, RAM, Storage
- ...

Purpose:

- Office use requires optical drive
- ...

Ex. 5, Task 1

Create an Alloy model for a scenario of your choice. The senario must make sense, i.e., not a sig A ... sig B example, and it needs to be different from the examples in the lecture.

- Declare at least 4 signatures each with at least 2 fields.
- Use inheritance between signatures at least once.
- Define at least 2 facts and 2 predicates.
- Add two run commands to your model, one unsatisfiable and one that returns at least 2 instances.

Ex. 7, Task 4

Create an SMV module to encode the moves of a knight on an 8x8 chess board. The knight always starts at coordinate (0, 1).
Knights can move as illustrated below ...

Fig. 2. Excerpts from exercise worksheets. The complete worksheets and task descriptions are available from https://github.com/fm4se/exercises/.

teaching workload, it provided significant value to the courses for students and instructors.

Angulo and Aktunc [2] have shared their insights on utilizing GitHub as an effective teaching tool for programming courses. They observed that GH offers advantages to educators by providing a collaborative classroom environment. Glassey [9] surveyed eight tools designed to ease the technical challenges of Git and GH in an educational setting, analyzing their general, technical, and pedagogical aspects. In a separate case study [10], Haaranen and Lehtinen shared their extensive experience teaching web software development using Git from both the instructors' and learners' perspectives. Kertész [16] found that students in an Operating Systems class preferred GHC's collaborative platform. They reported that the benefits of collaboration include learning from mistakes, receiving faster assistance from peers, mastering a common development platform, and improving decision-making through diverse solutions.

GitHub Classroom is also utilized to teach formal methods. To illustrate, Divasón and Romero [7] utilized GHC to submit their students' exercises in formal verification teaching, while Rozier [26] employed GHC to teach applied formal methods. In both instances, all homework was distributed and collected through GitHub Classroom. However, the authors provided no further details regarding their experience with the platform.

3.2   Autograding and Feedback

Automated grading of exercises has been a common feature in computer science courses for a long time [13, 22]. Messer et al. [22] conducted a systematic review and report a focus mainly on OOP-languages and the correctness aspect. They report limitations in the quality of feedback and imposing limits on showing creativity. Haldeman et al. [12] proposed a methodology for extending autograders to provide meaningful feedback by collecting and analyzing exercise submissions and generating hints that can be used in future semesters. Later, they proposed a framework called CSF^2 [11], which provides formative feedback on programming exercises. Although these systems and frameworks are related, they primarily focus on programming languages and may be challenging to adapt in the context of formal methods.

3.3   Automatic Exercise Generation

The automatic generation of programming exercises and the concept of autograding have piqued interest. Sovietov [29] developed a general scheme for generating programming exercises in a Python language course. This scheme can produce intricate exercises that demand complex solutions from students. Tiam-Lee and Sumi [30] presented a method for generating customized, entry-level programming exercises, while Sarsa et al. [27] utilized large language models to generate programming exercises automatically. The content generated was assessed qualitatively and quantitatively, mostly novel and sensible. Tscherter [31] proposed the Exorciser system for generating automatic exercises for a theory of computation class, e.g., regular languages and Markov algorithms. Exorciser provides feedback and visualization. Ábrahám et al. [1] discussed challenges in designing and generating automatic exercises for satisfiability checking and setting quality criteria for exercise generation.

4   Migration of Exercises to GitHub Classroom

GitHub Classroom (GHC) is an educational tool that helps teachers use Git and GitHub in their classes. Instructors create an assignment linked to a template repository, generating an invitation URL for students [2,32]. When a student accepts, an individual repository is created in the same GitHub organization. All worksheets are available as GitHub template projects from https://github.com/fm4se/exercises (Apache 2.0 license).

4.1   GHC Setup and Exercises

To transition from our traditional worksheets to GHC, we have implemented a task repository (template for replicated student repositories) for each worksheet. In our case task repositories are Java projects using the Gradle Build Tool[1]. Each repository's README.md (the *worksheet*) contains a comprehensive task breakdown and additional resources like videos and examples.

We have leveraged GitHub Actions to create an automated testing system and Continuous Integration (CI) workflow for evaluating student submissions. Our tailored workflow installs required dependencies and binaries, e.g., limboole [4], Z3 [23], nuXmv [5], related to each exercise. For some exercises, the workflow generates unique tasks (as in Fig. 2 (Ex. 3, Task 2)). We have employed a Community Action with a point bar[2] to display the exercise completion rate.

Further GHC Features and Limitations. We allowed students to complete the exercises in pairs. GHC enables group assignments, and students are responsible for joining their own groups. However, joining the wrong group can make it challenging for the instructor to locate the correct repository for grading purposes [32]. To avoid this confusion, we provide administration rights to the students for adding their team members to the repositories.

GHC enables linking learning management systems (LMS) to import student rosters [2]. We did not link the systems to avoid data protection concerns and instead asked students to submit their repository links on our LMS (Moodle).

The free GitHub plan only offers a monthly allocation of 2,000 GitHub Action minutes per organization. This may limit the execution of test cases and report generation. Our setup did not get anywhere near these limits.

---

[1] Gradle Build Tool https://gradle.org/.
[2] Available at https://github.com/marketplace/actions/points-bar.

4.2   Tests and Automated Feedback

Providing feedback is essential for students to comprehend the quality, depth, and relevance expected of their work [28]. It offers specific information regarding their learning growth, which is a great motivator. Moreover, it encourages students to reflect on their work.

Automated feedback offers immediate evaluation of student work, empowering them to improve before the final assessment. Traditional submit-and-mark exercises have limitations, including long wait times, no opportunity for improvement after submission, and less relevant feedback. Descriptive feedback helps students understand what went wrong and why, improving their work.

Safeguarding Test Integrity. Task completion is measured by test cases (as shown in Table 1). Providing these as part of the task repository exposes them to tampering, e.g., a student could modify test cases to always pass. To address this concern, we isolated the test cases within a secure cloud-based environment. During the GitHub Actions evaluation, we retrieved these test cases, integrated them into the student code, and executed the tests.[3]

Task 4: Knight Moves

Test	Status	Reason
start (0,1)	☑ Passed	-
next (2,2)	✘ Failed	⚠ RuntimeException
reach (7,7)	✘ Failed	Starting at (0,1) the knight can reach (7,7)

Fig. 3. Excerpt of a feedback report of Ex. 7, Task 4 from Fig. 2

Generating Feedback Report. When running JUnit tests with Gradle, an XML report is generated with detailed information about the test results. However, this XML report is not very intuitive to read for students. As a solution, we have parsed this XML report to create a comprehensive and easily understandable feedback report in a markdown format for students. An excerpt of the feedback report for Ex.7, Task 4 from Fig. 2 is presented in Fig. 3.

Local Feedback and Report Generation. GHC generates a feedback report for every push to the repository. However, students may want to generate and inspect feedback locally. We used Gradle and JUnit to execute the test cases, and a Python script to create the feedback report. Given a local installation and the required binaries of the FM tools, students may run the test cases in their IDE or by using the command `./gradlew test`. Given a Python installation they may further generate the feedback report via `./gradlew generateReport`.

---

[3] Recently, GHC introduced *protected file paths*, which flag changes in essential files, including tests, providing an out-of-the-box solution to detect tampering.

```
1 @Test
2 void checkNumberOfFieldsPerSignature() {
3   Module world = getModule(Tasks.task_1);   //parse with Alloy APIs
4   for (Sig s : world.getAllSigs()) {  //iterate over signatures
5     assertTrue(s.getFields().size() >= 2,  //check min num. fields
6       "Number of fields is less than 2 in signature " + s.label);
7   }
8 }
```

Listing 1. Example of using Alloy API for checking one criterion of the free-response task Ex. 5, Task 1 from Fig. 2: "... signatures each with at least 2 fields"

Managing Feedback Report. Automated feedback is pushed to a separate branch of each repository to avoid the task of merging changes from the primary branch. Figure 3 presents an excerpt of the automated feedback report for Ex.7, Task 4 from Fig. 2. The report comprises three columns: test, status, and reason for failures. The first test checks if the knight starts at (0,1) (as required by the task) and passes. The next test assesses whether the knight can move to (2,2) in the next step, but it fails due to a runtime error (this could be due to parsing errors). Lastly, the third test fails because the knight should be able to reach (7,7) starting from (0,1), which is not correctly implemented.

4.3 Migration Challenges

We reused most tasks from our worksheets with minor adjustments. We now list some challenges for moving from worksheet-based to GHC exercises. We indicate for each challenge whether it relates to specification tasks (*spec*), implementation tasks (*impl*), or to provide meaningful feedback (*feedback*).

Free-Response Questions. (challenges. *spec*, *feedback*) Our worksheets included many free-response questions, e.g., "Provide a satisfiable formula with at least 3 three variables and at least 3 different operators" or Ex. 5, Task 1 from Fig. 2, to evaluate higher-order thinking [18] and suppress plagiarism. To evaluate answers in GHC, we had to implement ad-hoc parsers or use APIs where available, e.g., for parsing and inspecting Alloy modules as in Listing 1. Both are not trivial and add overhead when setting exercises.

Problem Encoding. (challenges: *spec*, *impl*) Many tasks require coming up with suitable domain concept formalizations, e.g., PC component types in SMT (as in Fig. 2 (Ex. 3, Task 3)) can be encoded as integer constants or as datatypes (with their own pros/cons). We believe that this challenge/freedom is very important, but it makes an automated assessment of submissions challenging. We developed some minimal specification interfaces, e.g., variables for the budget and purpose of PCs, to construct and assess scenarios independent of the chosen encoding. The test case in Listing 2 only uses this specification interface (variables **purpose** and **budget**) to check a possible PC configuration. In general,

```
1  @Test
2  void testPurposeOffice1(){
3    assertTrue(Z3Utils.isSatForConstraints(code,   //check possible PC config
4      "(assert (= purpose office))\n(assert (= budget 283))"),
5      "For office purpose, 283 Euro is enough.");   //feedback if failed
6  }
```

Listing 2. Example of using a specification interface (variables `purpose` and `budget`) for checking a possible PC configuration of Ex. 3, Task 3 from Fig. 2

```
1  @Test
2  public void testTask1a() throws IOException, InterruptedException {
3    String testSpec = code + "\nLTLSPEC\n!(F(knight[0]=7&knight[1]=7));";
4    String res = NuxmvExecutor.runNuxmv(testSpec);   //run nuXmv with spec
5    assertTrue(res.contains(   //check nuXmv output for expected result
6      "!( F (knight[0] = 7 & knight[1] = 7)) is false"),
7      "Starting at (0,1) the knight can reach (7,7)");   //feedback if failed
8  }
```

Listing 3. Example of verifying additional LTL specifications on possible knight moves to check the user-defined transition relation of Ex. 7, Task 4 from Fig. 2

finding a balance of what variables to provide is non-trivial and specific to each task.

Solutions in Test Cases. (challenges: *spec, feedback*) One way to test a correct encoding of constraints is by checking semantic entailment or equivalence with a solution constraint, e.g., employed in Alloy4Fun [19]. To improve feedback, we decided to publish test cases for inspection to our students. A simple check of constraint equivalence was no longer possible as it would reveal the solution constraints (as in Fig. 2 (Ex. 3, Task 3)). Instead, we constructed multiple satisfiable and contradicting constraints (e.g., Listing 2 or Listing 3) for analysis and feedback generation.

Scenario Encoding. For effective analysis, we had to develop individual methods to encode and check scenarios (test cases of specifications) in SAT, Alloy, SMT, and nuXmv, e.g., to check transition relations of an SMV module (as in Fig. 2 (Ex. 7, Task 4)), our tests add LTLSPEC constraints and check these (see Listing 3). This requires creative extensions of specifications with constraints and the interpretation of the generated output of tools, as not all provide APIs.

Submission Format. (challenges: *spec*) A further challenge was to find a canonical format for collecting and submitting specification documents. Previously, students submitted PDFs, but specifications from these are not easily extracted. We used the interactive online editor Formal Methods Playground[4]

---
[4] Available from https://play.formal-methods.net.

(FMP). While separate, external storage of specifications makes the submission less self-contained, the students were used to FMP from lectures and links embedded in lecture slides. We also provided templates to get started, where necessary.

False Positives. (challenges: *impl, feedback*) Some test cases pass when no constraints are generated; thus, some scores might go down when providing correct but partial solutions. This risks confusion or discouragement of students. We handled these cases by adding explicit explanations on the worksheets.

Task Dependencies. (challenges: *impl, feedback*) Our implementation exercises were mini-projects where later tasks might depend on earlier ones, e.g., finding dead features in a Feature Model requires correct encoding of Feature Models. When grading manually, one could trace failures of later tasks to errors in earlier ones and award partial marks. In our setup, this was not possible. We indicated dependencies on the worksheets to make students aware of the grading limitation.

Order and Number of Test Cases. (challenges: *feedback*) Although not relevant from a correctness point of view, we suspect that the order of test cases has an impact on delivering feedback. Typically, students solve exercises incrementally, and we ordered test cases to provide early, positive feedback, e.g., assessing adequate numbers of signatures in an Alloy module before analyzing its predicates. Our implementation tasks already contained (limited) JUnit tests in the first iteration to confirm that students were going in the right direction. We significantly extended the number of test cases and assertions for feedback generation and correctness checking. Still, some students asked for covering additional edge cases (see Sect. 5).

Long-Running Test Cases and Exceptions. (challenges: *feedback*) Runtime errors, memory leaks, and exceptions thrown during testing may fail or interrupt test cases. We provide feedback on the type of exception, but understanding why the exception occurred without analyzing the code is challenging. Heavy computation, like infinite loops, further complicates the feedback process. We addressed long-running test cases by introducing timeouts, and we later (from Ex. 4) provided all test cases for local analysis of exceptions.

Additional Challenges. (challenges: *spec, impl, feedback*) (1) We were not able to automatically generate meaningful feedback for text-based questions, e.g., "Explain the provided nuXmv counter-example". (2) For limboole and nuXmv, where we did not have parsers nor APIs, our analyses depend on output of the tools and are thus fragile to changes, e.g., tool updates. (3) Some students exploited the limited number of test cases, e.g., the SMT encoding of PC components in one submission handles some edge cases of budget values by returning an empty result.

4.4   On Test Creation Efforts

Some of the challenges listed above similarly apply to general programming exercises regardless of the inclusion of Formal Methods, e.g., *False positives* or *Task dependencies*. However, others are clearly specific, e.g., *Problem encoding* or *Scenario encoding*.

One challenge we found during the later inspection of submissions is related to the intended use of FM tools in implementation exercises (Ex. 2, 4, and 6). One pair of students solved most analysis tasks by a translation to SMT; however, they did not find a suitable encoding for one edge case and handled it on the Java level (exploiting a limited number of test cases). While handling parts of the problem in Java and others in SMT might be elegant, this was not the intended learning outcome here. In general, our setup of the exercises would make it very difficult to assess whether solutions are computed in Java or SMT.

Table 1 summarizes the total number of test cases and assertions for each exercise and the number of test cases made available to students. Each test case offers feedback through a varied set of assertions (often, later assertions are not meaningful without earlier ones). The median number of test cases is 21, while the median number of assertions is 33.

Table 1. Number of test cases, assertions (different feedback texts), and test cases published to students per exercise; after exercise 3, all test cases were published

Exercises	No. of Test Cases	No. of Assertions	Published Test Cases
1	31	56	0
2	23	52	4
3	21	48	0
4	9	15	9
5	31	30	31
6	15	26	15
7	10	33	10

For exercises 1 and 3, no test cases were provided to students. Similarly, exercise 2 only had 4 test cases out of 23 available to students. However, after reviewing responses in student surveys, we provided all the test cases for exercises 4 to 7 to students. This potentially reduced their difficulties and helped them work better, as reflected in Figs. 5 and 7. Obviously, without providing test cases, we would have avoided challenge *Solutions in test cases*. However, we believe that the more transparent feedback outweighs the necessary adjustments.

One concern regarding the up-front availability of all test cases is that students could tweak their specifications and code just enough to make all test cases pass rather than develop complete solutions to the given tasks. An alternative is to split the test cases into two sets, one for feedback and one for marking. We leave a deeper evaluation of this to future work.

We notice that the preparation of exercises for GHC involves a significant investment of resources. Designing appropriate tasks and worksheets to cover taught materials and intended learning outcomes remained the same with or without GHC. Designing reasonable feedback and grading through test cases can be very challenging and difficult to get right, as all possible reasonable solution attempts by students should be taken into account. For us, this investment was paid back by reducing the time to inspect and grade homework submissions manually (only necessary in edge cases). However, low-quality or inadequate test cases could diminish this return on investment.

Finally, collusion and students passing on solutions over the years harm the reuse of the exercises for future iterations. We tried to mitigate this by including free-response questions and individually generated tasks in the specification exercises. For the implementation exercises, we employed plagiarism-checking tools.

4.5   Benefits and Alternatives to GHC

As discussed in Sect. 4.1 we did not use all features of GHC. Still, for our case, it provided several benefits: automated replication of task repositories for students, automated execution of test cases and report generation, user management of student accounts, and available infrastructure. However, our exercise materials are not tied to GHC. The setup may be replicated on any other infrastructure that provides repositories and continuous integration.

5   Surveys and Evaluation

Over the past two years (WS22-23 and WS23-24), we have been conducting the Formal Methods for Software Engineering module. In the first iteration, a total of 86 students enrolled (24 groups submitted the first exercise), while in the second iteration, 44 students enrolled (11 groups submitted the first exercise)[5]. Figure 4 illustrates the number of groups per exercise. We observed that almost half of the groups dropped out in both semesters over all the exercises submitted. In WS22-23, submissions went from 24 groups to only 13 groups for the last exercise. Similarly, in WS23-24, we received 11 submissions for the first exercise but only 5 for the last exercise.

5.1   Surveys in WS23-24

We now report data from voluntary paper-based surveys conducted among students on the days after the submission of exercises in WS23-24. We conducted these surveys to gain insight into the effectiveness of implementing GitHub Classroom with automated feedback. The surveys were designed to evaluate the ease

---

[5] Enrolling is an informal act for obtaining access to teaching materials and the numbers of initial submissions are more reliable indicators for participation.

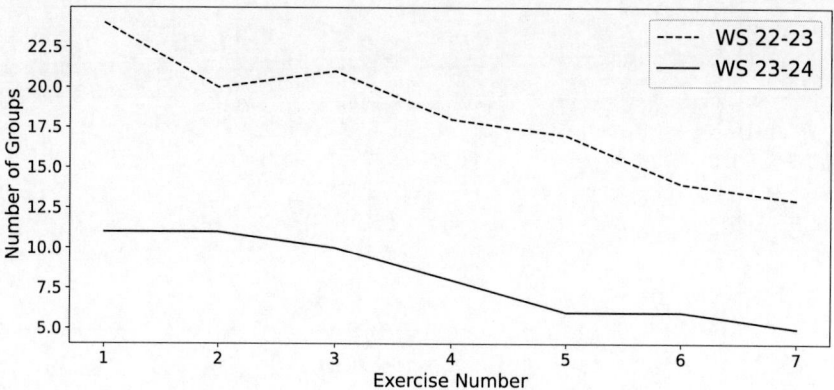

Fig. 4. Submission per exercise for the Winter semesters of 2022-23 and 2023-24

of using GitHub for submitting exercises, the quality of automated feedback, and whether it contributed to a perceived overall improvement of student work or additional workload. Additionally, we gathered suggestions from students to refine the submission process and enhance the quality of the exercises.

Figure 5 illustrates the declared level of difficulty experienced by students while using GitHub to work on and submit exercises. Notably, most students, including those from the Digital Engineering program (mixed background from engineering disciplines), found using GitHub to be easy. Additionally, we observed that the students grew more comfortable with the platform over time. In the initial two exercises, the median difficulty level reported by students was "neutral" to "easy", whereas for the subsequent exercises, the median level ranged from "easy" to "very easy". It should be highlighted that the number of students participating in the exercises and surveys decreased over time.

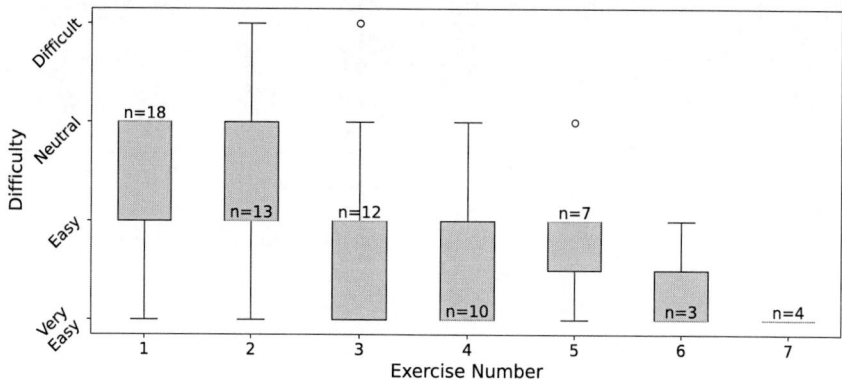

Fig. 5. The difficulty of using GitHub for completing the exercises

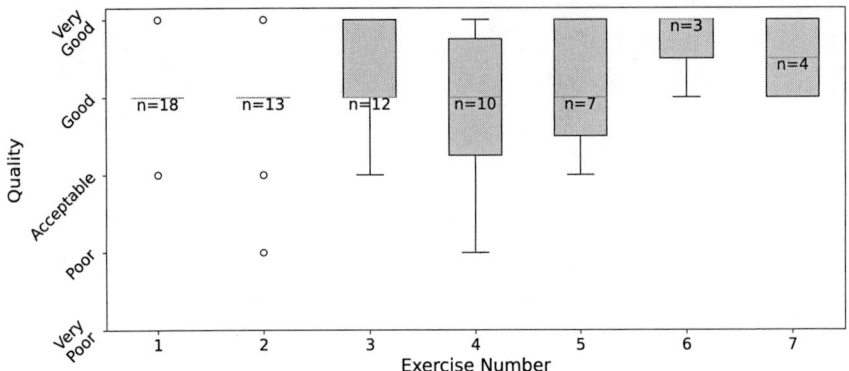

Fig. 6. The quality of the automated feedback for the exercises

Figure 6 shows the perceived quality of the automated feedback for the exercises. At first, we relied solely on the feedback given by GitHub Classroom through the GitHub Actions logs for the initial exercise. Unfortunately, but not surprisingly, the students found interpreting and comprehending the feedback challenging. As a result, we began automatically extracting feedback from test reports and aggregating it in tabular format starting from exercise 3. We observed that the quality of the feedback remained "good" for the first five exercises and even improved for the final two.

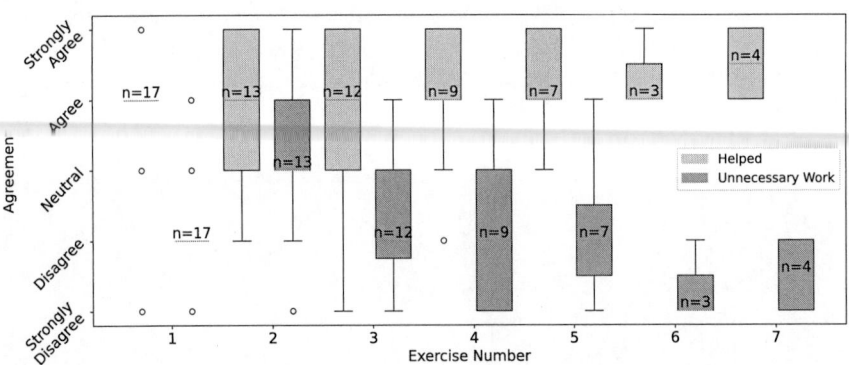

Fig. 7. Agreement on automated feedback being helpful (blue) or unnecessarily causing more work (red) (Color figure online)

The data presented in Fig. 7 (blue) illustrates how students agree with the usefulness of automated feedback in improving their work. At the same time, Fig. 7 (red) highlights their level of agreement on whether generated feedback caused a higher workload. The majority of students believe that automated feedback has been beneficial in enhancing their work across all exercises, as reflected by the median. However, there were some instances of disagreement among students in exercises 2 and 3 (test cases not available to students), with 75% remaining neutral or strongly agreeing. Nonetheless, automated feedback was found to be helpful in improving their work in subsequent exercises. In terms of additional workload, students generally disagreed that feedback added to their workload, except for exercise 2, where the median was different.

In addition to the Likert scale questions above, our survey also includes open-ended prompts asking about challenges and suggestions. We analyzed the textual responses and grouped them into two categories: automated feedback and GH/GHC. These categories align with the themes presented in Figs. 5-7.

Automated Feedback. Students expressed appreciation for automated feedback, citing benefits like improved work quality, bug resolution, and better understanding of different scenarios:

- *The automated feedback was very important to evaluate in our cases it was very nice to know what improvements can be made in the code further.*
- *The feedback really helps with the process of completing and understanding the tasks. If a problem is encountered, the feedback helps in identifying the topic of concept that needs to be revised for completion.*
- *Continuous feedback on each statement helped me compare and understand the assignment better.*

However, they also identified areas for improvement, such as wanting more detailed feedback and the need to check all corner cases:

- *The automated tests didn't test for multiple components of the some category, which should not be possible.*
- *Provide more information on why the test case has failed and also the exact errors.*
- *Maybe include test cases or in this case the LTLSPECS in the playground template for easy access.*

GH/GHC. Overall, students had a positive experience using GitHub and GitHub Classroom despite facing some challenges including group submissions and not having all the test cases locally (communicated verbally in class), which were provided later for exercises 4–7:

- *The assignment submissions on GitHub really helped me know what is going wrong with any work looking at the feedback.*
- *There can be better infrastructure assignments for group submissions of the assignments.*

5.2   Threats to Validity

Likert scales are well-known to be subject to various forms of response bias [17], thus, one has to be careful with conclusions solely based on these. Another potential bias in our data is the decline in the number of answers (n in Figs. 5-7). While all figures show a positive trend, we cannot make conclusions about whether the quality of worksheets improved, the students became more experienced with GHC or the feedback, or whether simply the more experienced or confident students remained participating.

6   Conclusion

We have presented a brief overview of our module, Formal Methods for Software Engineering, and its two exercise types (specification and implementation). We identified challenges in migrating from traditional submissions to GitHub Classroom (GHC) with automated grading and feedback report creation. Student surveys indicate low difficulty of using GHC, good quality and helpfulness of automated feedback, and low overhead for students.

References

1. Ábrahám, E., Nalbach, J., Promies, V.: Automated exercise generation for satisfiability checking. In: Dubois, C., San Pietro, P. (eds.) Formal Methods Teaching. FMTea 2023. LNCS, vol. 13962, pp. 1–16. Springer, Cham (2023). https://doi.org/10.1007/978-3-031-27534-0_1
2. Angulo, M.A., Aktunc, O.: Using GitHub as a teaching tool for programming courses. In: 2018 Gulf Southwest Section Conference Proceedings, p. 31594. ASEE Conferences. https://doi.org/10.18260/1-2-370.620-31594, http://peer.asee.org/31594
3. Baier, D., Beyer, D., Friedberger, K.: JavaSMT 3: interacting with SMT solvers in java. In: Silva, A., Leino, K.R.M. (eds.) CAV 2021. LNCS, vol. 12760, pp. 195–208. Springer, Cham (2021). https://doi.org/10.1007/978-3-030-81688-9_9
4. Biere, A.: Limboole (2012). https://fmv.jku.at/limboole/. Accessed July 2024
5. Cavada, R., et al.: The NUXMV symbolic model checker. In: Biere, A., Bloem, R. (eds.) CAV 2014. LNCS, vol. 8559, pp. 334–342. Springer, Cham (2014). https://doi.org/10.1007/978-3-319-08867-9_22
6. Davies, J., Simpson, A., Martin, A.: Teaching formal methods in context. In: Dean, C.N., Boute, R.T. (eds.) TFM 2004. LNCS, vol. 3294, pp. 185–202. Springer, Heidelberg (2004). https://doi.org/10.1007/978-3-540-30472-2_12
7. Divasón, J., Romero, A.: Using Krakatoa for teaching formal verification of java programs. In: Dongol, B., Petre, L., Smith, G. (eds.) FMTea 2019. LNCS, vol. 11758, pp. 37–51. Springer, Cham (2019). https://doi.org/10.1007/978-3-030-32441-4_3
8. Dubois, C., Prevosto, V., Burel, G.: Teaching formal methods to future engineers. In: Dongol, B., Petre, L., Smith, G. (eds.) FMTea 2019. LNCS, vol. 11758, pp. 69–80. Springer, Cham (2019). https://doi.org/10.1007/978-3-030-32441-4_5

9. Glassey, R.: Adopting Git/Github within teaching: a survey of tool support. In: CompEd 2019, pp. 143–149. ACM (2019). https://doi.org/10.1145/3300115. 3309518
10. Haaranen, L., Lehtinen, T.: Teaching Git on the side: version control system as a course platform. In: ITICSE 2015, pp. 87–92. ACM (2015). https://doi.org/10. 1145/2729094.2742608
11. Haldeman, G., Babeş-Vroman, M., Tjang, A., Nguyen, T.D.: CSF: formative feedback in autograding **21**(3), 1–3 (2021). https://doi.org/10.1145/3445983
12. Haldeman, G., et al.: Providing meaningful feedback for autograding of programming assignments. In: SIGCSE 2018, pp. 278–283. ACM. https://doi.org/10.1145/ 3159450.3159502
13. Hollingsworth, J.: Automatic graders for programming classes. Commun. ACM **3**(10), 528–529 (1960). https://doi.org/10.1145/367415.367422
14. Hsing, C., Gennarelli, V.: Using GitHub in the classroom predicts student learning outcomes and classroom experiences: findings from a survey of students and teachers. In: SIGCSE 2019, pp. 672–678. SIGCSE 2019, ACM (2019). https://doi. org/10.1145/3287324.3287460
15. Jackson, D.: Alloy: a language and tool for exploring software designs. Commun. ACM **62**(9), 66–76 (2019). https://doi.org/10.1145/3338843
16. Kertész, C.Z.: Using GitHub in the classroom - a collaborative learning experience. In: SIITME 2015, pp. 381–386 (2015). https://doi.org/10.1109/SIITME. 2015.7342358
17. Liu, M., Harbaugh, A.G., Harring, J.R., Hancock, G.R.: The effect of extreme response and non-extreme response styles on testing measurement invariance. Front. Psychol. **8**, 227387 (2017)
18. Luxton-Reilly, A., Denny, P., Plimmer, B., Bertinshaw, D.J.: Supporting student-generated free-response questions. In: ITiCSE 2011, pp. 153–157. ACM (2011). https://doi.org/10.1145/1999747.1999792
19. Macedo, N., et al.: Experiences on teaching alloy with an automated assessment platform. Sci. Comput. Program. **211**, 102690 (2021). https://doi.org/10.1016/J. SCICO.2021.102690
20. Marques-Silva, J., Malik, S.: Propositional SAT Solving. In: Handbook of Model Checking, pp. 247–275. Springer, Cham (2018). https://doi.org/10.1007/978-3-319-10575-8_9
21. Mendonça, M., Wasowski, A., Czarnecki, K.: SAT-based analysis of feature models is easy. In: Muthig, D., McGregor, J.D. (eds.) SPLC 2009. ACM International Conference Proceeding Series, vol. 446, pp. 231–240. ACM (2009). https://dl.acm. org/citation.cfm?id=1753267
22. Messer, M., Brown, N.C.C., Kölling, M., Shi, M.: Automated grading and feedback tools for programming education: a systematic review. ACM Trans. Comput. Educ. **24**(1), 10:1–10:43 (2024). https://doi.org/10.1145/3636515
23. de Moura, L., Bjørner, N.: Z3: an efficient SMT solver. In: Ramakrishnan, C.R., Rehof, J. (eds.) TACAS 2008. LNCS, vol. 4963, pp. 337–340. Springer, Heidelberg (2008). https://doi.org/10.1007/978-3-540-78800-3_24
24. Pelletier, F.J.: Seventy-five problems for testing automatic theorem provers. J. Autom. Reason. **2**(2), 191–216 (1986). https://doi.org/10.1007/BF02432151
25. Pnueli, A.: The temporal logic of programs. In: SFCS 1977, pp. 46–57. IEEE Computer Society (1977). https://doi.org/10.1109/SFCS.1977.32
26. Rozier, K.Y.: On teaching applied formal methods in aerospace engineering. In: Dongol, B., Petre, L., Smith, G. (eds.) FMTea 2019. LNCS, vol. 11758, pp. 111–131. Springer, Cham (2019). https://doi.org/10.1007/978-3-030-32441-4_8

27. Sarsa, S., Denny, P., Hellas, A., Leinonen, J.: Automatic generation of programming exercises and code explanations using large language models. In: ICER 2022 - Volume 1, pp. 27–43. ACM (2022). https://doi.org/10.1145/3501385.3543957
28. Shute, V.J.: Focus on formative feedback. Rev. Educ. Res. **78**(1), 153–189 (2008)
29. Sovietov, P.: Automatic Generation of Programming Exercises. In: TELE 2021, pp. 111–114. IEEE (2021). https://doi.org/10.1109/TELE52840.2021.9482762
30. Tiam-Lee, T.J.Z., Sumi, K.: Procedural generation of programming exercises with guides based on the student's emotion. In: SMC 2018, pp. 1465–1470. IEEE (2018). https://doi.org/10.1109/SMC.2018.00255
31. Tscherter, V.: Exorciser: automatic generation and interactive grading of structured excercises in the theory of computation. Ph.D. thesis, ETH Zurich, Zürich, Switzerland (2004). https://doi.org/10.3929/ETHZ-A-004830877
32. Tu, Y.C., et al.: GitHub in the classroom: lessons learnt. In: ACE 2022, pp. 163–172. ACM. https://doi.org/10.1145/3511861.3511879

Teaching Through Practice: Advanced Static Analysis with LiSA

Luca Negrini[1], Vincenzo Arceri[2(✉)], Luca Olivieri[1], Agostino Cortesi[1],
and Pietro Ferrara[1]

[1] Ca' Foscari University of Venice, Venice, Italy
{luca.negrini,luca.olivieri,cortesi,pietro.ferrara}@unive.it
[2] University of Parma, Parma, Italy
vincenzo.arceri@unipr.it

Abstract. Nowadays, ready-to-use libraries and code generation are often used to streamline and speed up the software development process. The resulting programs are thus a collection of different modules that cooperate: proving their safety and reliability is increasingly complex, requiring sound formal techniques, such as static program analysis. However, while teaching static analysis to master's or PhD students, the predominant focus on theoretical concepts often leaves limited space for students to engage with the practical aspects of implementing static analyses and is limited to developing elementary ones. In this paper, we show how the infrastructure offered by LiSA can be exploited to learn how to implement advanced static analyses, such as string and relational numerical analyses, just focusing on their distinctive aspects. This would help to narrow the gap between theoretical and practical contents in static analysis courses, bringing the learning experience beyond the rudimentary implementation of static analyses to more sophisticated applications.

1 Introduction

Static analysis based on formal methods requires a non-trivial theoretical background and development skills. Traditional static analysis courses based on formal methods at the master's or doctoral level often give priority to mathematical-theoretical concepts, leaving limited room for students to actively engage with the practical aspects of implementing static analyses. Moreover, the design and implementation of new static analyses require building an infrastructure providing several basic building blocks, depending on the mathematical theory used. Therefore, developing even a toy static analyzer from scratch is a significant effort that could discourage students and teachers.

The goal of this paper is to show how we can exploit LiSA[10,15] to reduce practical and development efforts for the building of static analysis solutions based on abstract interpretation [6,7], allowing students to put their hands on more advanced static analyses based on formal methods, and providing ready-to-use components to develop their custom implementations. In particular, we choose to discuss and present two of the static analyses: the prefix abstract

E. Sekerinski and L. Ribeiro (Eds.): FMTea 2024, LNCS 14939, pp. 43–57, 2024.
https://doi.org/10.1007/978-3-031-71379-8_3

domain [5], and the pentagon abstract domain [12], showing how LiSA allows us to focus just on the implementation of the peculiar aspects of a static analysis of interest to come up with a complete and ready-to-run analysis. We then discuss how we use LiSA to teach practical static analysis in a Computer Science master course.

Paper Structure. Section 2 provides an informal and high-level introduction to static analysis by abstract interpretation. Section 3 describes the key components of LiSA. Section 4 reports the implementation details in LiSA of two advanced abstract domains, namely prefix abstract domain and pentagon abstract domain. Section 5 discusses how we integrated LiSA in the master course we teach. Finally, Sect. 6 concludes and illustrates how we intend to use LiSA to improve the teaching experience within static analysis courses.

2   Static Analysis by Abstract Interpretation

Static analysis is a technique used for inspecting program properties without *concretely* executing the program. Examples of these properties may be whether the program terminates, which program variables are constants, or whether a program contains safety and security issues (e.g., buffer overflows [9], injection vulnerabilities [21], data leaks [11]).

For static analysis to *guarantee* the presence (or absence) of code properties, bugs, and vulnerabilities, one must adopt an approach based on a formal methods framework. Among these frameworks, a notable example is certainly abstract interpretation [6,7]. Abstract interpretation is a theoretical framework that provides a systematic way to correctly approximate program behaviors and reason about some properties of the program of interest on such approximation. One of the fundamental concepts of abstract interpretation is the notion of *abstract domain,* which provides an abstraction of the concrete program states as a set of *abstract values.* The goal of an abstract domain, typically modeled as a (complete) lattice [7], is to capture just the relevant aspects of program behavior while discarding details irrelevant to the analysis of interest. Regarding teaching, abstract interpretation principles require a non-trivial theoretical background, such as the notions of lattices, domains, fix-point theorems, and Galois connections [6]. Theoretical concepts thus tend to dominate the available time for the course, allowing only a shallow (and often not practical) exploration of simple abstractions.

The classic candidate to teach abstract domains is the *sign domain* [7], where numerical variables are abstracted to capture their sign, i.e., positive (Pos), negative (Neg), or zero (Zero). It is often chosen for its simplicity and intuitiveness. In particular, the sign abstract domain is depicted by its Hasse diagram reported in Fig. 1a, where the partial order between abstract values, the least upper bound, and the greatest lower bound are highlighted. For instance, Fig. 1b shows what the sign analysis infers on a simple code fragment.

At line 1, variables a and b are abstracted to the abstract value Pos because the concrete assigned values 5 and 7 are positive integer numbers. At line 3,

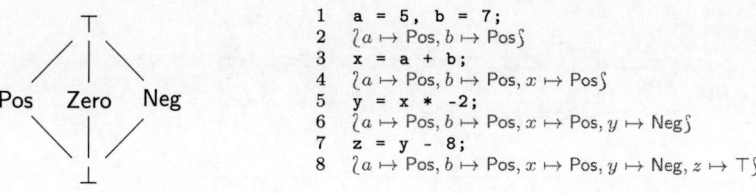

```
1   a = 5, b = 7;
2   ⎨a ↦ Pos, b ↦ Pos⎬
3   x = a + b;
4   ⎨a ↦ Pos, b ↦ Pos, x ↦ Pos⎬
5   y = x * -2;
6   ⎨a ↦ Pos, b ↦ Pos, x ↦ Pos, y ↦ Neg⎬
7   z = y - 8;
8   ⎨a ↦ Pos, b ↦ Pos, x ↦ Pos, y ↦ Neg, z ↦ T⎬
```

(a) Hasse diagram of signs.   (b) The sign domain in action.

Fig. 1. The sign abstract domain.

in order to infer the correct sign of the variable x, it is necessary to define the so-called *abstract semantics* of the assignment and the sum operator. In general, once we have defined how our concrete values (e.g., integers) are abstracted into an abstract domain (e.g., sign), it is also necessary to define the abstract semantics of the operations in a program of interests, i.e., how each operator affects the abstract states represented by the abstract domain. Back to our example, the abstract semantics of the sign domain for the sum operator corresponds to the classical sign rules (e.g., Pos + Pos = Pos). Hence, the analysis infers that the sign of x is Pos. At line 7, the abstract semantics of the minus operator between two Pos abstract values returns the top abstract value ⊤, and it assigns it to z, meaning that the analysis is not able to determine the sign for z, because we are reasoning on abstract values (e.g., signs) and not on the concrete ones (e.g., integers).

Non-relational and Relational Abstract Domains. The sign domain shown above can be seen as *"the Hello World of static analysis"*, being typically the first (and most common) numerical domain to be used to introduce static analysis by abstract interpretation, for instance, in master courses or PhD schools; this abstract domain lends itself well for this being a *non-relational abstract domain*, i.e., an abstract domain that does not explicitly model relationships between different variables, treating each variable independently. In contrast, *relational abstract domains*, such as pentagons [12], octagons [13,20] or convex polyhedra [3,8], also capture relationships between different variables.

Generally speaking, relational abstract domains offer higher precision than non-relational ones, but they are also more complex to define and require additional computational efforts to track and maintain the relationships between program variables.

3   LiSA

LiSA is a modular framework for developing static analyzers based on the abstract interpretation theory. LiSA was born as a tool for research purposes (e.g., [16–18]). However, its modular infrastructure also enabled us to use it to teach static analysis by abstract interpretation. The high-level analysis process

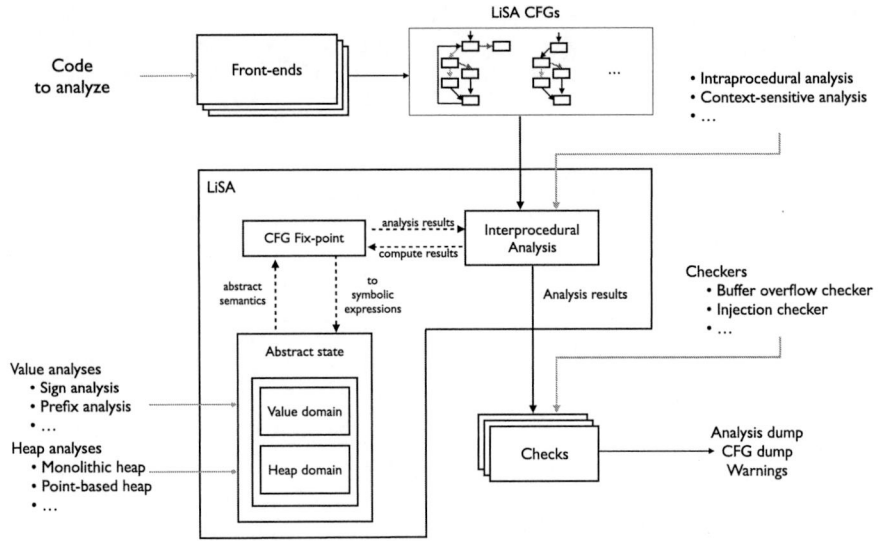

Fig. 2. LiSA overall execution.

of LiSA is reported in Fig. 2. LiSA analyzes control-flow graphs (CFGs) [1], a representation that expresses the control structure of the code using graph notation. In particular, LiSA uses a general design for CFGs, where statements do not have predefined semantics; instead, users of the framework can define custom statement instances implementing language-specific semantic functions, enabling the analysis of a wide range of programming languages and the development of multilanguage analyses. The analysis infrastructure is partitioned into three main areas: call evaluation, memory modeling, and value analysis. Each area corresponds to a configurable analysis component that operates agnostically concerning how the others are implemented. The analysis begins in the *Interprocedural Analysis*, which executes a program-wide fixpoint by computing each individual CFG's fixpoint. Whenever a call is encountered, the computation of its result is delegated back to the *Interprocedural Analysis*. Instead, non-calling statements are decomposed into a sequence of atomic operations, called *symbolic expressions*, each with precise semantics that the abstract domains can interpret. Memory-dealing expressions are handled by the *Heap Domain*, tracking their effect and rewriting them as abstract identifiers representing possible memory locations. Finally, the *Value Domain* tracks properties about variables (either program variables or abstract identifiers) and computes invariants for each program point. At the end of the analysis, results can be inspected through *Checks*, which are program visitors that can access the computed invariants and that can use them to warnings about points in the program where, for example, a property of interest holds. Code parsing logic and the definition of language-specific statements are provided by *Frontends*, which can also provide implementations for LiSA's components. These constitute effective static analyzers for individual languages that can be combined to obtain multilanguage analyses. Several frontends have already been implemented, with new ones in the works. This paper and our courses use a frontend for IMP: a simple object-oriented language built for testing and demonstration. IMP is feature-rich enough to let the students practice with realistic static analysis without over-complicating the semantics abstraction. The IMP frontend is shipped with LiSA.[1] Students can use the various implementations of each analysis component provided within LiSA, thus focusing only on the one they are experimenting with.

Key Components for Implementing an Abstract Domain. In abstract interpretation courses, students usually become familiar with non-relational numeric domains, such as sign and interval [7]. We provide here the necessary notions not only to implement those domains but also other non-relational ones possibly dealing with non-numeric data, as well as relational one.

A (relational or non-relational) domain tracking values of variables must implement the `ValueDomain` interface. This requires providing all of the lattice operators of the domain (e.g., lub, partial order, widening) as well as *semantic transformers* to be invoked during fixpoints to track the semantics of each symbolic expression processed. `ValueDomain` implementers must provide, among other ones, (i) an `assign` method, invoked to store the result of an expression into a variable, and (ii) a `smallStepSemantics` method, invoked when the effect of a non-assigning symbolic expression is to be evaluated. Such methods transform the domain instance that receives the call into a new one according to the received expression(s) effects.

When coding a non-relational domain, some aspects of the implementation are independent of the domain itself. There is always a mapping from variables to instances of the domain's values, lattice operators are defined through functional lifting over such map, and the domain only evolves with assignments after recursively evaluating the right-hand side to a domain's value. To simplify such implementation tasks, LiSA ships with a `ValueDomain` implementation named `ValueEnvironment`, parametric to a `NonRelationalValueDomain` (NRVD for short). Such domain (i) is effectively a map from variables to instances of the NRVD, (ii) uses functional lifting for lattice operators, delegating to NRVD for the values of the mapping, (ii) has a `smallStepSemantics` implementation that is a no-op, (iii) has an `assign` implementation that evaluates the right-hand side to an instance of the NRVD and maps it to the target variable. Thus, an NRVD is mainly required to provide (i) lattice operator *for individual values* (e.g., between signs, instead of the whole mapping), and (ii) an evaluation logic for expressions. As the recursive visiting of a symbolic expression is independent of the NRVD of choice, LiSA also ships a subclass of it called `BaseNonRelationalValueDomain` (BNRVD for short) providing such logic. Implementers of BNRVD thus only have to provide (other than lattice operators) the evaluation logic of individual expressions given the value of their arguments.

---

[1] The IMP specification is available at https://lisa-analyzer.github.io/imp/.

4   Analyses Implementation with LiSA

4.1   An Example String Analysis: the Prefix Domain

String analysis focuses on tracking program properties concerning strings. In the context of string static analysis by abstract interpretation, several abstract domains have been proposed, each with different properties of interest and complexity: prefix, suffix, char inclusion, bricks, and string graph abstract domains [5], automata- and regex-based abstractions [4,14,22], and relational string analyses [2], are just a few examples. While LiSA implements most of the aforementioned abstract domains, in the rest of this paper, we show the teaching-related peculiar aspects of implementing the prefix abstract domain.[2] As the name suggests, the prefix abstract domain is a non-relational abstraction that keeps track of the prefix of string variables. Consider the following code fragment to give a flavor of how the prefix domain works.

```
1    if (x == 2) {
2        s = "javaSE";
3        ⟨s ↦ javaSE*⟩
4    } else {
5        s = "javascript";
6        ⟨s ↦ javascript*⟩
7    }
8    ⟨s ↦ java*⟩
```

Supposing that the variable x has a statically unknown value, the value of the Boolean guard at line 1 is not determined, and to infer the prefix abstraction for the variable s at line 8 (i.e., at the end of if-statement), both branches must be taken into account by the analysis; the true-branch abstracts the value of s to $javaSE*$, that is the (concrete) value of s starts with the string "$javaSE$". Similarly, the false-branch abstracts the value of s to $javascript*$. At line 8, the analysis infers that the abstract value of s is $java*$ by applying the prefix's least upper bound between the two abstract values.

Prefix Abstract Domain Implementation. In the following, we report the implementation of the prefix abstract domain with LiSA.[3] In particular, thanks to LiSA, one only needs to implement the peculiar parts of the abstract domain to come up with a ready-to-use analysis. Concerning the prefix abstract domain, the key points are that: (i) it must be a *non-relational* domain, (ii) it needs a mechanism to keep track of abstract prefixes, and (iii) the lattice-related operators and abstract semantics must be implemented.

We can define the domain as a class called Prefix, implementing the BNRVD interface (Fig. 3). The class is characterized by the prefix field (Fig. 3 at line 3) to keep track of string prefix abstract values. The value of this field is set during the construction of an abstract value element (Fig. 3 at line 5).

---

[2] Full details about the definition and formalization of the prefix abstract domain can be found at [5].

[3] Full code available at https://github.com/lisa-analyzer/lisa/blob/master/lisa/lisa-analyses/src/main/java/it/unive/lisa/analysis/string/Prefix.java.

```
1   public class Prefix implements BaseNonRelationalValueDomain<Prefix>
2   {
3     private final String prefix;
4     // [...]
5     public Prefix(String prefix) { this.prefix = prefix; }
6     // [...]
7   }
```

Fig. 3. Implementation of the class `Prefix`.

Concerning lattice-based operations such as least upper bound and partial order operation, LiSA provides a base implementation for lattice abstract domains called `BaseLattice`, also implemented by BNRVD, that provides a base implementation for lattice operations refactoring behaviors that are common to most of the non-relational abstract domains. For instance, the least upper bound between the top (resp. bottom) element and any other abstract value, always returns top (resp. the other abstract value). Since `Prefix` implements BNRVD (and in turn `BaseLattice`), the least upper bound operator is implemented through `lubAux`, reported in Fig. 4, that can ignore cases where both `this` and `other` are the same abstract value, or are equal to bottom or top (being handled in `BaseLattice`). `lubAux` computes the longest common prefix between `this` and `other` (Fig. 4 at lines 3, where the definition of `longestCommonPrefix` is left implicit). If the common prefix is not empty, a new prefix abstract element is returned; otherwise, the top element is returned (Fig. 4 at line 4). Partial order and the greatest lower bound can be implemented similarly.

```
1   @Override
2   public Prefix lubAux(Prefix other) {
3     String result = longestCommonPrefix(this.prefix, other.prefix);
4     return result.isEmpty() ? TOP : new Prefix(result.toString());
5   }
```

Fig. 4. Least upper bound operator of `Prefix`.

Finally, BNRVD provides callbacks for evaluating the abstract semantics, one for each LiSA symbolic expression type. The following shows the abstract semantics for non-null constant values and binary string domain-specific expressions. To define the abstract semantics for string literals, we need to implement the `evalNonNullConstant` method, reported in Fig. 5, that returns a new prefix abstract value a value if the constant is a string, or the top element for the non-string constant values (e.g., integers).

Concerning binary string expressions, we instead implement the `evalBinary-Expression` method, reported in Fig. 6, that takes two abstract values `left` and `right` and the binary operator `op` that applies to them. Among the LiSA string symbolic expressions, the prefix abstract domain can infer a non-top prefix abstract value only for string concatenation (Fig. 6 at line 4). In contrast, for all the others, it returns top. Specifically, the prefix abstract value of `left` concatenated with `right`, returns `left`.

```
1   @Override
2   public Prefix evalNonNullConstant(Constant constant) {
3     if (constant.getValue() instanceof String) {
4       String str = (String) constant.getValue();
5       if (!str.isEmpty()) { return new Prefix(str); }
6     }
7     return TOP;
8   }
```

Fig. 5. Abstract semantic implementation of non-null constant values.

```
1   @Override
2   public Prefix evalBinaryExpression(BinaryOperator op, Prefix left,
3       Prefix right) {
4     if (op instanceof StringConcat) { return left; }
5     return TOP;
6   }
```

Fig. 6. Abstract semantic implementation of binary expressions.

4.2   An Example Relational Analyses: the Pentagon Domain

In Sect. 2, we highlighted that relational analyses are usually more complex abstractions to be designed and implemented. In terms of teaching, these may not be trivial for students first approaching static analysis. Here, we detail the implementation in LiSA of the Pentagon abstract domain [12]. This weakly relational numeric abstract domain is relatively simple compared to others but contains all the basic notions to understand relational domains. The Pentagon domain captures relations of the form $x \in [a, b] \land x < y$ and consists of two sub-domains: a non-relational interval abstraction ($x \in [a, b]$) combined with a relational *strict upper bound* domain ($x < y$). The two components are combined via (an abstraction of the) *reduced product* [7], corresponding to a Cartesian product where the two domains mutually exchange information in order to refine each other.

```
1   ⁀x ↦ [0, +∞], y ↦ [0, +∞]⁁
2   if (x > y) {
3       ⁀x ↦ [0, +∞], y ↦ [0, +∞], x > y⁁
4       r = x - y
5       ⁀x ↦ [1, +∞], y ↦ [0, +∞], r ↦ [0, +∞], x > y⁁
6       assert(r >= 0)
7       ⁀x ↦ [1, +∞], y ↦ [0, +∞], r ↦ [0, +∞], x > y⁁
8   }
```

Fig. 7. Code fragment example taken from [12, Sect. 6.2.1].

Let us consider the code fragment reported in Fig. 7, where variables x and y are initially abstracted to $[0, +\infty]$. The interval abstract domain is not precise enough to prove the assertion at line 6 (intuitively, the assignment of r at line 4 yields the interval $[-\infty, +\infty]$). Instead, the Pentagon abstract domain also tracks strict relations between variables, as highlighted by the invariant related to x and y at line 3. Thus, exploiting the information x > y, we can refine x −

y, and in turn, the interval assigned to r, in $[0, +\infty]$;[4] this is enough to prove the assertion at line 6.

Pentagon Abstract Domain Implementation. Here, we report part of the implementation of the Pentagon abstract domain. The key points are that: (i) it must be a *relational* domain, (ii) it needs a mechanism to manage the information related to non-relational interval and strict upper bounds domains, (iii) the lattice-related operators and abstract semantics must be implemented.

```
1   public class Pentagon implements ValueDomain< Pentagon >,
2       BaseLattice <Pentagon> {
3       private final ValueEnvironment<Interval> intervals;
4       private final ValueEnvironment <UpperBounds > upperBounds;
5       // [...]
6   }
```

Fig. 8. Implementation of the class `Pentagon`.

Unlike the prefix domain implementation presented in Sect. 4.1, Pentagon is a relational domain and cannot exploit the BNRVD interface. The class `Pentagon` is reported in Fig. 8, implementing both the `ValueDomain` and `BaseLattice` interfaces. The domain is characterized by the field `interval`, keeping track of the interval of each variable, and the field `upperBounds`, keeping track of the relations of the form $x < y$ between two variables; the latter is implemented as an environment mapping each variable to a set of variables: for instance, $x \mapsto \{y, z\} \implies x < y \wedge x < z$. We will focus on how the two components of this domain can interact and exchange information, omitting implementation details of the components related to interval and upper bounds abstract domains.[5]

Being `Pentagon` a `ValueDomain` implementer, it must provide the implementation for the `smallStepSemantics` and `assign` methods. Concerning the former, the implementation is as simple as relying on the `smallStepSemantics` of the two sub-components of the domain, as reported in Fig. 9 at lines 1–4.

More attention is needed to implement the `assign` method, reported in Fig. 9 at lines 6–25, where refinement of the two sub-components of the Pentagon abstract domain may occur. The `assign` method takes an expression e that needs to be assigned to an identifier id (line 6). Lines 7–8 perform the assignment on the interval and the upper bounds abstractions calling their respective `assign` methods. Next, we check if it is possible to refine the abstraction: here, we discuss the refinement concerning assignments of the form $id = x - y$. If the assigned

---

[4] Formal specification of the Pentagon's abstract semantics for subtraction can be found in [12].

[5] Implementation of the interval and upper bound abstract domains can be found at https://github.com/lisa-analyzer/lisa/blob/master/lisa/lisa-analyses/src/main/java/it/unive/lisa/analysis/numeric/Interval.java and https://github.com/lisa-analyzer/lisa/blob/master/lisa/lisa-analyses/src/main/java/it/unive/lisa/analysis/numeric/UpperBounds.java.

```
1    public Pentagon smallStepSemantics(ValueExpression expression) {
2      return new Pentagon(intervals.smallStepSemantics(expression),
3             upperBounds.smallStepSemantics(expression));
4    }
5
6    public Pentagon assign(Identifier id, ValueExpression e) {
7      ValueEnvironment<UpperBounds> newBounds = upperBounds.assign(id, e);
8      ValueEnvironment<Interval> newIntvs = intervals.assign(id, e);
9      // refinement
10     if (e instanceof BinaryExpression) {
11       BinaryExpression be = (BinaryExpression) e;
12       if (be.getOperator() instanceof SubtractionOperator
13           && be.getLeft() instanceof Identifier
14           && be.getRight() instanceof Identifier) {
15         // id = x - y
16         Identifier x = (Identifier) be.getLeft();
17         Identifier y = (Identifier) be.getRight();
18         if (newBounds.getState(y).contains(x))
19           newIntvs = newIntvs.putState(id, newIntvs.getState(id).glb(
20             new Interval(MathNumber.ONE, MathNumber.PLUS_INFINITY)));
21       }
22       // [...]
23     }
24     return new Pentagon(newIntvs, newBounds).closure();
25   }
```

Fig. 9. Semantic transformers of `Pentagon`.

expression is in the form $x - y$ (lines 12–14), it is possible to refine the interval abstraction for variable *id* if the upper bounds domain knows that $y < x$. This is checked at line 18, and if so, the interval for `id` is refined, restricting it to $[1, +\infty]$, applying the greatest lower bound operator (lines 19–20). Finally, the result is returned at line 33 after applying the transitive closure (left implicit) on the abstract value.

4.3   Run LiSA

Once the desired abstract domain is developed, it is ready to analyze programs with the following fragment.

```
1    Program p = IMPFrontend.processFile(filePath);
2    LiSAConfiguration conf = new DefaultConfiguration();
3    conf.workdir = "output";
4    conf.analysisGraphs = GraphType.DOT;
5    conf.abstractState = new SimpleAbstractState<>(
6      new MonolithicHeap(),
7      new Pentagon(),
8      new TypeEnvironment<>(new InferredTypes()));
9    new LiSA(conf).run(p);
```

Line 1 uses the IMP frontend to parse an IMP program, returning a LiSA `Program`. Lines 2–8 build a LiSA configuration, setting the working directory, how to dump the analysis results, and the desired static analysis; we choose to use the `SimpleAbstractState` class, requiring a heap domain (monolithic heap), a value domain (`Pentagons`) and a type domain (non-relational type environment), that come with LiSA. Finally, line 9 runs the analysis, producing untyped petagons_tests::common_code_pattern_01(petagons_tests* this, untyped x, untyped y, untyped r)

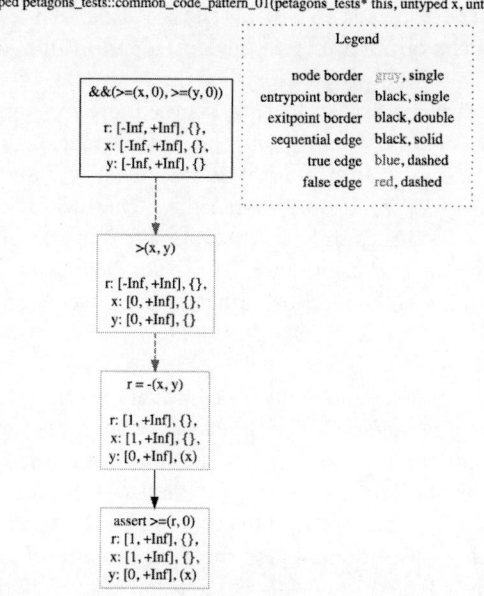

Fig. 10. LiSA dump for the program reported in Fig. 7.

the graph reported in Fig. 10 with the pentagon analysis results.[6] Note that to run the `Prefix` analysis instead, it is sufficient to change line 7 to `new ValueEnvironment<>(new Prefix())`.

5   Our Teaching Experience with LiSA

LiSA has been used for the last four years for the practical part of the Computer Science master course *Software Correctness, Security and Reliability* at University Ca' Foscari of Venice. Every year, the course features 30 to 50 students with a standard Computer Science background (in particular, set theory and object-oriented programming are given as prerequisites). Students learn lattice theory and abstract interpretation during the course, paired with practical experience using LiSA.

Course Structure. During the first three years, students took part in three lectures exclusively focused on LiSA: a first introductory lecture giving a full overview of the structure and the analysis process (Sect. 3), a second one with a focus on dataflow analyses, and a third one focusing on simple non-relational numeric analyses [10]. The second and third lectures also featured a live coding session, going through the relevant classes and interfaces and tackling the intricacies of the implementations. After each coding session, students were assigned

---

[6] For space limitations, we have omitted details about the heap and type analyses.

coding tasks covering the topics seen during the lecture, with a one-week deadline. Students also had the opportunity to develop an additional analysis, following a reference paper, as a final project for the course instead of taking the theoretical exam. The fourth year instead featured three more lectures on LiSA, each including live coding: one on advanced non-relational numeric analysis (typically the interval domain [7]), one on information flow analysis [19] and string analysis (Sect. 4.1), and a final one on relational numeric analyses (Sect. 4.2). Tasks were also given for these new lectures, and the theoretical exam was removed in favor of a mandatory project on the usage of LiSA to analyze real-world programs. The lectures were initially delivered through online classes (mandatory due to Covid-19 restrictions) but moved to in-person lectures during the last two years.

Settings for Coding. Coding sessions and task implementations aimed at targeting increasingly challenging scenarios without overwhelming the students with technical details or complex language features. Programs targeted were thus written in IMP to (i) exploit the IMP-frontend provided by LiSA instead of defining one from scratch, and (ii) have a simplified semantics to reason upon. Moreover, students used built-in components for the *Interprocedural Analysis* and the *Heap Domain*, allowing them to focus specifically on the peculiar aspects of an abstract domain while ignoring more complex and out-of-scope reasoning still needed to make the analysis run.

Tasks and Exams Evaluation. Tasks and final exams were graded from 0 to 10, assigning points for (i) the correct usage of LiSA as seen during classes, (ii) the correctness of the implementation w.r.t. the formal definition of the domain, and (iii) the correctness of the results produced over some simple yet expressive code snippets. Grades for the third point were assigned automatically, while the first two required manual inspection of the code submitted. Since the tasks were mandatory, all students actively engaged in solving them. Each year, marks gradually grew with each task, starting from an average of 6 out of 10 in the first task to a 7.5 in the second one. For the fourth year, having three more lectures proved beneficial: students reached an average of 8.8 marks on the fifth task. In the first three years, the number of students choosing the practical project varied between 10% to 40%, showing that most students were still hesitant to engage in a more challenging analysis development. This was the key motivation for the increase in practical lectures during the fourth year. At the time of writing, final projects for the fourth year are still in progress, so we do not have reports on the impact of such additional lectures on the students' understanding of static analysis.

Impact on Students' Careers. During all four years, we registered an increase in the number of students asking for a master's thesis on static analysis w.r.t. the number of requests in the previous editions of the course, with 2–3 students per year. Moreover, in the last three years, we disseminated LiSA through dedicated invited seminars inside other static analysis courses in two neighboring universities. Currently, the researchers and professors from such universities are proposing bachelor and master theses projects related to LiSA, leading to additional eight students working with it, overall.

6   Conclusion

In this paper, we presented and described how LiSA can be exploited in a static analysis course to put the student's hands on two examples of advanced static analyses: a string analysis and a relational numeric analysis.

The adoption of LiSA into our educational curriculum has already yielded significant benefits in our previous courses on static analyses, allowing students to engage in the implementation of sophisticated static analyses. Furthermore, the ready-to-use components gave students a complete overview of all the essential components necessary to build static analyses via abstract interpretation while experimenting with the analyses they were tasked with. Additionally, LiSA also helps students develop a deeper understanding of the more practical applications of static analysis, especially during final course projects.

Given this success, noted by the positive feedback from the students, the next semester will see a heavier emphasis on using LiSA in our courses to bridge further the gap between theory and practice in static analysis education.

Acknowledgements. Work partially supported by Bando di Ateneo per la Ricerca 2022, funded by University of Parma, (MUR_DM737_2022_FIL_ PRO-GETTI_B_ARCERI_COFIN, CUP: D91B21005370003), SERICS (PE00000014, CUP H73C2200089001), and iNEST (ECS00000043 - CUP H43C22000540006) projects funded by PNRR NextGeneration EU.

References

1. Allen, F.E.: Control Flow Analysis. In: Proceedings of a Symposium on Compiler Optimization, p. 1–19. Association for Computing Machinery, New York, NY, USA (1970). https://doi.org/10.1145/800028.808479
2. Arceri, V., Olliaro, M., Cortesi, A., Ferrara, P.: Relational string abstract domains. In: Finkbeiner, B., Wies, T. (eds.) VMCAI 2022. LNCS, vol. 13182, pp. 20–42. Springer, Cham (2022). https://doi.org/10.1007/978-3-030-94583-1_2
3. Becchi, A., Zaffanella, E.: PPlite: zero-overhead encoding of NNC polyhedra. Inf. Comput. **275**, 104620 (2020). https://doi.org/10.1016/J.IC.2020.104620
4. Christensen, A.S., Møller, A., Schwartzbach, M.I.: Precise analysis of string expressions. In: Cousot, R. (ed.) SAS 2003. LNCS, vol. 2694, pp. 1–18. Springer, Heidelberg (2003). https://doi.org/10.1007/3-540-44898-5_1
5. Costantini, G., Ferrara, P., Cortesi, A.: A suite of abstract domains for static analysis of string values. Softw. Pract. Exp. **45**(2), 245–287 (2015). https://doi.org/10.1002/SPE.2218
6. Cousot, P.: Principles of Abstract Interpretation. MIT Press (2021)
7. Cousot, P., Cousot, R.: Abstract interpretation: a unified lattice model for static analysis of programs by construction or approximation of fixpoints. In: Graham, R.M., Harrison, M.A., Sethi, R. (eds.) Conference Record of the Fourth ACM Symposium on Principles of Programming Languages, Los Angeles, California,

USA, January 1977, pp. 238–252. ACM (1977). https://doi.org/10.1145/512950.512973

8. Cousot, P., Halbwachs, N.: Automatic discovery of linear restraints among variables of a program. In: Aho, A.V., Zilles, S.N., Szymanski, T.G. (eds.) Conference Record of the Fifth Annual ACM Symposium on Principles of Programming Languages, Tucson, Arizona, USA, January 1978, pp. 84–96. ACM Press (1978). https://doi.org/10.1145/512760.512770

9. Cowan, C., Wagle, F., Pu, C., Beattie, S., Walpole, J.: Buffer overflows: attacks and defenses for the vulnerability of the decade. In: Proceedings DARPA Information Survivability Conference and Exposition. DISCEX'00, vol. 2, vol. 2, pp. 119–129 (2000). https://doi.org/10.1109/DISCEX.2000.821514

10. Ferrara, P., Negrini, L., Arceri, V., Cortesi, A.: Static analysis for dummies: experiencing lisa. In: Do, L.N.Q., Urban, C. (eds.) SOAP@PLDI 2021: Proceedings of the 10th ACM SIGPLAN International Workshop on the State of the Art in Program Analysis, Virtual Event, Canada, 22 June, 2021, pp. 1–6. ACM (2021). https://doi.org/10.1145/3460946.3464316

11. Ferrara, P., Olivieri, L., Spoto, F.: Static privacy analysis by flow reconstruction of tainted data. Int. J. Softw. Eng. Know. Eng. **31**(07), 973–1016 (2021). https://doi.org/10.1142/S0218194021500303

12. Logozzo, F., Fähndrich, M.: Pentagons: a weakly relational abstract domain for the efficient validation of array accesses. Sci. Comput. Program. **75**(9), 796–807 (2010). https://doi.org/10.1016/J.SCICO.2009.04.004

13. Miné, A.: The octagon abstract domain. High. Order Symb. Comput. **19**(1), 31–100 (2006). https://doi.org/10.1007/S10990-006-8609-1

14. Negrini, L., Arceri, V., Ferrara, P., Cortesi, A.: Twinning automata and regular expressions for string static analysis. In: Henglein, F., Shoham, S., Vizel, Y. (eds.) VMCAI 2021. LNCS, vol. 12597, pp. 267–290. Springer, Cham (2021). https://doi.org/10.1007/978-3-030-67067-2_13

15. Negrini, L., Ferrara, P., Arceri, V., Cortesi, A.: LiSA: a Generic Framework for Multilanguage Static Analysis. In: Arceri, V., Cortesi, A., Ferrara, P., Olliaro, M. (eds.) Challenges of Software Verification. Intelligent Systems Reference Library, vol. 238, pp. 19–42 Springer, Singapore (2020). https://doi.org/10.1007/978-981-19-9601-6_2

16. Negrini, L., Shabadi, G., Urban, C.: Static analysis of data transformations in Jupyter notebooks. In: Ferrara, P., Hadarean, L. (eds.) Proceedings of the 12th ACM SIGPLAN International Workshop on the State Of the Art in Program Analysis, SOAP 2023, Orlando, FL, USA, 17 June 2023, pp. 8–13. ACM (2023). https://doi.org/10.1145/3589250.3596145

17. Olivieri, L., Jensen, T.P., Negrini, L., Spoto, F.: MichelsonLiSA: a static analyzer for Tezos. In: IEEE International Conference on Pervasive Computing and Communications Workshops and other Affiliated Events, PerCom Workshops 2023, Atlanta, GA, USA, 13-17 March 2023, pp. 80–85. IEEE (2023). https://doi.org/10.1109/PERCOMWORKSHOPS56833.2023.10150247

18. Olivieri, L., et al.: Information flow analysis for detecting non-determinism in blockchain. In: Ali, K., Salvaneschi, G. (eds.) 37th European Conference on Object-Oriented Programming, ECOOP 2023, 17-21 July 2023, Seattle, Washington, United States. LIPIcs, vol. 263, pp. 23:1–23:25. Schloss Dagstuhl - Leibniz-Zentrum für Informatik (2023). https://doi.org/10.4230/LIPICS.ECOOP.2023.23

19. Sabelfeld, A., Myers, A.C.: Language-based information-flow security. IEEE J. Sel. A. Commun. **21**(1), 5–19 (2006)

20. Schwarz, M., Seidl, H.: Octagons revisited - elegant proofs and simplified algorithms. In: Hermenegildo, M.V., Morales, J.F. (eds.) Static Analysis - 30th International Symposium, SAS 2023, Cascais, Portugal, October 22-24, 2023, Proceedings. Lecture Notes in Computer Science, vol. 14284, pp. 485–507. Springer (2023). https://doi.org/10.1007/978-3-031-44245-2_21
21. Spoto, F., et al.: Static identification of injection attacks in Java. ACM Trans. Program. Lang. Syst. **41**(3) (2019). https://doi.org/10.1145/3332371
22. Veanes, M.: Applications of symbolic finite automata. In: Konstantinidis, S. (ed.) CIAA 2013. LNCS, vol. 7982, pp. 16–23. Springer, Heidelberg (2013). https://doi.org/10.1007/978-3-642-39274-0_3

Teaching Formal Methods for 10 Years: Reflections on theories, tools, materials, and communities

Gustavo Carvalho[✉][iD]

Centro de Informática, Universidade Federal de Pernambuco, Recife 50.740-560, Brazil

`ghpc@cin.ufpe.br`

Abstract. Software is virtually everywhere. As a software-based society, thinking about correctness is an important skill of the present and future software and system developers. Since Formal Methods play an important role in achieving high-trustworthiness levels, they have been directly or indirectly part of typical Computer Science education. This paper discusses and summarises impressions taken from a 10-years experience on teaching Formal Methods to undergraduate students enrolled at Computer Science and Computer Engineering courses in Brazil. At first, the topic was taught using Z and CSP# as reference languages. Over the years, for a number of reasons, the chosen languages changed to a combination of Event-B and CSP_M and, at the present time, it is being taught using B. Our reflections are based on the following four perspectives: theories, tools, materials, and communities. We build on these reflections to highlight aspects that should be taken into account when designing Formal Methods modules as part of a Computer Science curriculum.

Keywords: teaching reflections · formal methods · Z language · Event-B · B language · Communicating Sequential Processes

1 Introduction

While driving, flying, travelling, listening to music, playing games, doing exercises, studying, teaching, working, buying and selling goods, reading, writing, and communicating with each other, we use software. Software is virtually everywhere. As a software-based society, it is at least annoying when software and systems do not function as expected. More than just irritating, malfunctioning systems can be dangerous, as we should have already learned from past experiences. Therefore, Computer Scientists, Computer Engineers, and other professionals empowered with the challenges of developing software and systems should be trained to think about correctness.

Among the disciplines that contribute to achieving high-trustworthiness levels, Formal Methods play an important role. According to FME, a worldwide association of researchers and practitioners in Formal Methods, *formal methods are mathematical approaches to software and system development which support the rigorous specification, design and verification of computer systems*[1]. Within the particular context of cybersecurity, the importance of Formal Methods has been recently acknowledged by The White House, which now considers these methods as part of the agenda of its National Cybersecurity Strategy [13].

Formal Methods education has been directly or indirectly part of typical Computer Science curricula for decades. Authors of a recent paper advocate a prominent role of Formal Methods in the ACM CS 2023 curriculum; particularly, emphasising the importance of Formal Methods thinking [3]. According to them, it provides the necessary rigour when reasoning about correctness. They go further and argue that this is a subject every computer scientist needs to know.

An online directory of Formal Methods courses is made available by the FME Teaching Committee[2]. We would summarise the approaches of these courses into the following four categories: (i) a unified overview, (ii) a fragmented overview, (iii) an in-depth account, and (iv) some combination of the previous ones. A unified overview generally chooses a single formal method to illustrate fundamental concepts such as, formal specification, refinement strategies, and formal verification. Differently, a fragmented overview takes into account different formalisms to illustrate such concepts, typically exploring various aspects of systems (e.g., functional behaviours, concurrency, time properties, probabilistic behaviours, among others). Opposed to the two previous approaches, some courses delve into the details of a specific Formal Method (e.g., understanding the details of a particular proof assistant). In other situations, the course begins with an overview and, then, follows an in-depth approach.

In Brazil, teaching Formal Methods has been part of many Computer Science courses. In older curricula, this used to be a mandatory module; whereas, in newer ones, it commonly appears as an optional one. We also have an important presence in Formal Methods research, with prominent researchers being part of international communities, taking part in large cooperation projects, in addition to acting as reviewers and editors of important publication venues. The main Brazilian symposium on Formal Methods is SBMF: an acronym for *Simpósio Brasileiro de Métodos Formais* (i.e., Brazilian Symposium on Formal Methods). The "Brazilian" in its name means that the event happens every year in some Brazilian city. Nevertheless, it is an international venue attended by researchers and practitioners from all over the world. In 2024, the event goes for its 27th edition[3]; its proceedings have been published by Springer in the form of Lecture Notes in Computer Science (LNCS).

In this paper, we discuss and summarise our impressions taken from teaching Formal Methods for 10 years to undergraduate students enrolled at Computer Science and Computer Engineering courses in Brazil. At first, the subject was

---

[1] Source: https://www.fmeurope.org/formalmethods/.

[2] Link: https://fme-teaching.github.io/courses/.

[3] Link: https://dblp.org/db/conf/sbmf/index.html.

taught using Z [15] and CSP# [12] as reference languages. Over the years, for a number of reasons, as discussed later, the chosen languages changed to a combination of Event-B [2] and CSP_M [10]. At the present time, it is being taught using B [1]. Our reflections are based on the following four perspectives: theories, tools, materials, and communities. We build on these reflections to highlight aspects that should be taken into account when designing Formal Methods modules as part of a Computer Science curriculum.

We structure this paper as follows. In Sect. 2, considering small toy examples, we briefly present the formalisms that have been taught. Afterwards, in Sect. 3, we detail our teaching experiences throughout these years. In Sect. 4, we build on this experience to provide a summary of reflections and recommendations. Finally, in Sect. 5, we address related work, and present our final remarks.

2   The Languages

Here, we provide a brief overview of the languages employed to teach Formal Methods. More attention is given to B, which is the one adopted nowadays. This way, all readers will be aware of some basic concepts of these languages.

The Z and CSP Languages. Z is based upon set theory and a first-order predicate calculus. A Z specification is structured according to several kinds of paragraphs. They define basic types, constraints, and schemas, among others. Schemas are used to describe the state of the system, and how it evolves.

The Communicating Sequential Processes (CSP) language focuses on specifying and reasoning about concurrent systems. The basic building blocks are events. They represent atomic actions that are performed by processes. A process is defined by combining events and referring to other processes. CSP offers operators specially designed to specify concurrent (synchronous and asynchronous) behaviour. There are two main dialects of CSP: CSP_M and CSP#. The former is a machine-readable version of CSP combined with a functional language. Differently, the latter combines CSP with an imperative and object-oriented language based on C#. For a comprehensive comparison of these dialects we refer the reader to [11].

The Event-B and B Languages. Event-B and classical B (or just B) are related languages. The former is suitable for describing event-driven reactive systems, whereas the latter focuses on developing imperative programs. Here, we use a classical puzzle (Tower of Hanoi) just to illustrate B; see Fig. 1.

An abstract `Machine` describes the behaviour of a system in mathematical terms. The machine `MHanoi` has a single variable: `hanoi`, whose type is a total function from 1..3 to the finite subsets of 1..3. It associates with each peg a set of disks – see how this variable is initialised. We assume that the minimum element of the set denotes the disk at the top. The operation `move` updates the system state. It has two parameters. The `from` peg must not be empty, and, if the `to`

```
1   /* Author: Gustavo Carvalho (ghpc@cin.ufpe.br) */
2  MACHINE
3       MHanoi
4  VARIABLES
5       hanoi
6  INVARIANT
7       hanoi : 1..3 --> FIN(1..3) & ...
8  INITIALISATION
9       hanoi := {1 |-> {1,2,3}, 2 |-> {}, 3 |-> {}}
10 OPERATIONS
11      move(from,to) =
12      PRE
13          from : dom(hanoi) & to : dom(hanoi) & from /= to
14          & hanoi(from) /= {}
15          & (hanoi(to) /= {} => min(hanoi(from)) < min(hanoi(to)))
16      THEN
17          hanoi := hanoi <+ { from |-> (hanoi(from) - {min(hanoi(from))}),
18                              to |-> (hanoi(to) \/ {min(hanoi(from))}) }
19      END
20 END
```

Fig. 1. B specification of the Tower of Hanoi puzzle.

peg is not empty, the disk that is being moved should be smaller than the disk at the top of the target peg. The state is updated by overloading the definition of `hanoi`: a disk is removed from the `from` peg (- denotes set subtraction), and added to the `to` peg (\/ denotes set union).

3   Teaching Experience

Throughout these 10 years, Formal Methods have been presented to students as a reinterpretation of Software Engineering, now with rigour. During the first classes, we discuss the area based on the perspective depicted in Fig. 2.

The central element is the formal model, which precisely describes the expected behaviour of the system. In some methods, this model is created by hand, whereas, in others, it is semi- or automatically generated from other classical Software Engineering artefacts, such as requirements, (in)formal models, or even code. Provided that there is a clear specification of the system, Formal Methods allow us to perform different tasks, grouped into four perspectives: animation, verification, translation, and visualisation.

Animation and visualisation play an important role in validating whether the correct model has been constructed. In the former, simulations are performed randomly or guided by users. Here, one needs to be familiar with the adopted formalism, since the formal model is directly manipulated during the simulation. Differently, in the latter, a graphical visualisation is built to enable validation by non-formal methods experts. Here, the focus is on domain specialists that interact indirectly with the model via synchronous views.

Verification encompasses providing (formal) guarantees that properties of interest hold. This can be accomplished with varying degrees of rigour and automation. One can benefit from (model-based) testing approaches, but also rely on more exhaustive approaches. Proof arguments can be put down on paper or be mechanised: interactively via proof assistants, or automatically with the aid of theorem provers and model checkers. Finally, translation addresses the challenges of translating the formal specification into other models or even code.

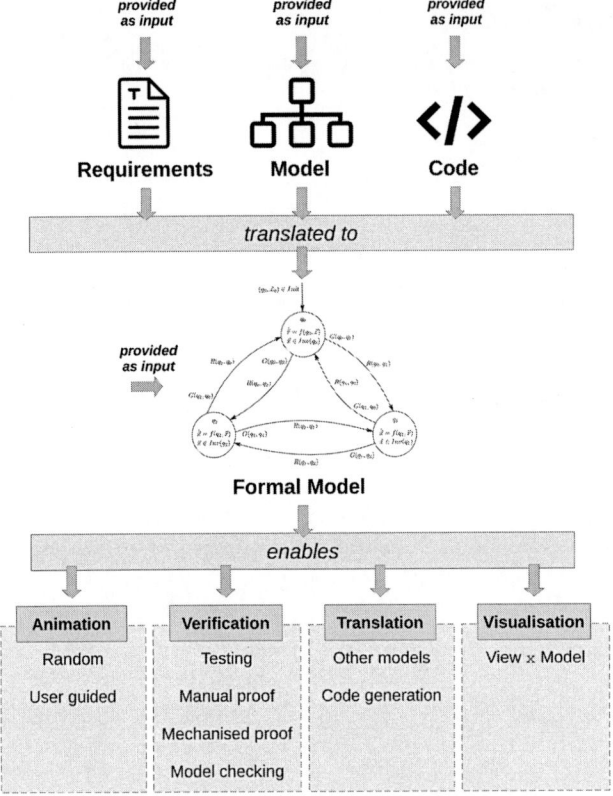

Fig. 2. An overview of Formal Methods.

The teaching experience considered here started in 2014, and continues to present days (2024). It is grouped into three distinct phases. In Phase 1, the reference languages for teaching Formal Methods were Z and CSP#. In Phase 2, Event-B and CSP_M were adopted instead. At the current phase (Phase 3), the subject is being taught using the B language. The teaching material for Phases 1–2 is no longer available online. Regarding Phase 3, it is publicly available at: https://sites.google.com/a/cin.ufpe.br/if721. In Sects. 3.1, 3.2, and 3.3, we provide more details about each of these three phases, respectively.

3.1   Phase 1: Teaching with Z and CSP#

Teaching Formal Methods started in 2014 as part of the curriculum of Computer Engineering at *Universidade de Pernambuco* (UPE[4]), which is a local public institution of higher education.

---

[4] Link: https://www.upe.br/.

General Information. In 2014, Formal Methods was a mandatory module taught at the 7th semester of the Computer Engineering course. The enrolled students typically had already attended Discrete Mathematics and Formal Logic, even though this was not formally required. In 2021, after a revision of the curriculum at UPE, the Formal Methods module became optional.

The main goal of the module is to present the students to various techniques associated with the formal development of systems. The syllabus covers topics such as formal notations, theorem provers, and model checking. The expectation is to provide an overview of these techniques, instead of getting into the details. Additionally, a review of Discrete Mathematics and Formal Logic is also part of the Formal Methods module to ensure a minimum knowledge of these subjects. We have 72 h scheduled for a mix of classroom and laboratory sessions. In Brazil, the academic year is composed of two semesters. Therefore, roughly speaking, a module lasts for one semester; for instance, 2014.2 denotes the module taught in the second semester of 2014.

At the end, the students should have developed the following capabilities: (i) recognise the importance of critical systems, (ii) differentiate traditional and formal approaches for developing software and systems, (iii) be capable of formally describing the expected behaviour and properties of a system from an informal specification, and (iv) be aware of classical techniques for formal verification, such as model checking and theorem proving.

Figure 3 shows the number of enrolments and fails over the years; Phase 1 comprises the semesters 2014.2 and 2015.1. When we started teaching Formal Methods in 2014.2, we had 17 students enrolled in the module. This is consistent with the number of student admissions per semester (40) and the number of students facing academic progress issues. In the following semester (2015.1), for some unclear reasons, the number of enrolments dropped to 5. Maybe the module was perceived as too difficult by the students, who decided to postpone it, since it is not a prerequisite to any other mandatory module (but only to optional ones). However, we do not have evidence to support this hypothesis.

Regarding the fail ratio, it was considered to be low. In 2014.2, 3 students failed for not reaching the minimum required mark. In 2015.1, 1 student failed for the same reason, whereas 1 student failed for not attending enough lectures.

Theories, Tools, Materials, and Communities. Influenced by our research experience, we started teaching the subject using a combination of Z and a dialect of CSP. To some extent, we were pursuing a combination of Z and CSP as proposed in Circus [14]. However, due to limitations of tool support, instead of using Circus, we focused on using the languages in isolation.

Concerning tools, The Z part was covered using the Community Z Tools – CZT[5], which augment the Eclipse IDE[6] to enable the development of Z specifications. A custom editor is provided with parsing and type checking features.

---

[5] Link: https://czt.sourceforge.net/.
[6] Link: https://eclipseide.org/.

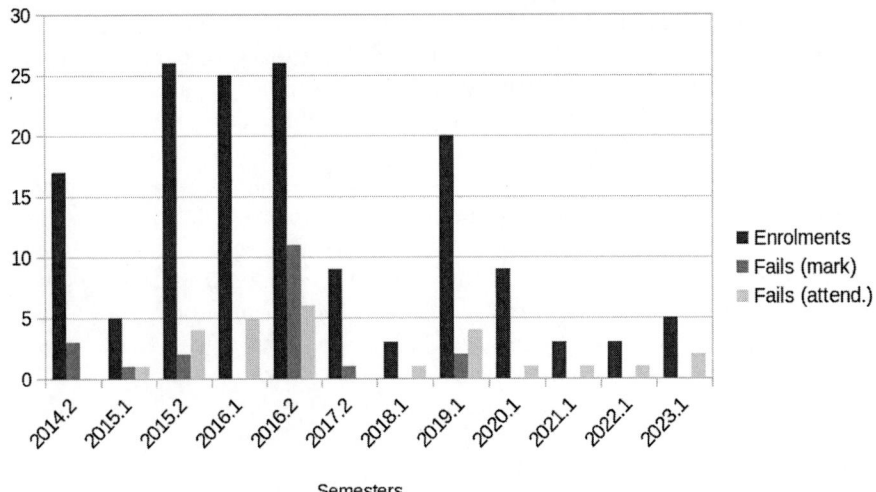

Fig. 3. Number of enrolments and fails (by mark and by attendance) over the years. Phase 1 = 2014.2 .. 2015.1. Phase 2 = 2015.2 .. 2016.2. Phase 3 = 2017.2 .. 2023.1.

Verification is enabled via integration with the Z-Eves theorem prover[7]. Nevertheless, during the module, there was not enough time to use Z-Eves.

Regarding CSP, we used the Process Analysis Toolkit – PAT[8], which deals with CSP#. The main tool for CSP_M is the Failures-Divergence Refinement checker (FDR)[9]. Tool support was our motivation for choosing CSP#. At that time, the second version of FDR (FDR2) had a GUI with limited features. PAT, besides an editing environment with the typically expected features (e.g., syntax highlighting, parsing, and type checking), provides a friendly simulator for interactively and visually simulating system behaviours. Verification is also automated by means of model checking. Particularly, one can carry out deadlock-freeness analysis, reachability analysis, linear temporal logic checking, and refinement checking. Some of these verification possibilities were explored in the module.

The book by Jim Woodcock and Jim Davies [15] was used as the main reading reference for Z. Concerning CSP#, we relied on a comprehensive tutorial embedded in the supporting tool. Nevertheless, the students used to complain about the limited number of introductory online resources and online communities. To mitigate the situation, we had many laboratory sessions in order to address practical doubts raised by them.

Teaching and Assessment Dynamics. As mentioned before, the teaching dynamics considered a mix of classroom and laboratory sessions. The module

---

[7] Link: https://czt.sourceforge.net/eclipse/zeves/.
[8] Link: https://pat.comp.nus.edu.sg/.
[9] Link: https://cocotec.io/fdr/.

was split into three parts. In the first one, which was by far the longest one, we focused on Z. CSP# was addressed in the second part. Finally, in the last part, we had seminars delivered by the students presenting an overview of other Formal Methods and tools, such as: Petri Nets, UPPAAL, PRISM, Isabelle/HOL, and Z3. Therefore, according to the categorisation discussed in Sect. 1, we would say that this module followed mainly a fragmented overview of Formal Methods.

The marks were determined by composing the assessment of a project (two parts – one devoted to Z, another to CSP#) and two written exams. It is important to say that all students worked on the same project, with detailed instructions provided to them.

Impressions. We summarise our impressions from this phase as follows.

- Although our underlying intention was to address the specification and verification of concurrent systems by integrating the modelling of states (Z) and concurrent behaviours (CSP), the students complained about the loose connection between the first and second parts of the module. Integrating Formal Methods carries a complexity that may be hard to grasp from the perspective of undergraduate students being first introduced to the topic.
- The students also used to complain about the time dedicated to teaching and learning CSP#. As they preferred its tool support, rather than the one associated with Z, they would like to spend more time working on the second part of the module.
- Specially for teaching purposes, it is essential to adopt user-friendly tools. Tools enable the combination of the theory and the practice, and help the students to see how the taught methods can be used in real contexts. As explained before, some choices we made were directly influenced by this. Ideally, these tools should be free for academic purposes and be multiplatform. Automated verification is also a plus, since some students may not be capable yet to develop more complicated proof arguments.
- The students appreciated that the teaching materials (i.e., the Z book, and the CSP# tutorial) were publicly available. However, they missed an online supporting community of practitioners. In general terms, we have academic publications that employ these languages in more advanced settings.
- The student seminars were rather shallow. They did not have the time and maturity to understand by themselves other methods and present a nice summary to the class.

3.2   Phase 2: Teaching with Event-B and CSP$_M$

By the end of 2015, the teaching format associated with Phase 1 started to progressively change to what we call Phase 2. First, we started teaching Formal Methods with Event-B, instead of Z. Later, we also changed from CSP# to CSP$_M$. Phase 2 concerns a teaching experience at UPE, as Phase 1.

General Information. The second phase is related to the same module (Formal Methods) of Phase 1. At that time, it was still mandatory at UPE. The syllabus and expectations were the same as described in Phase 1.

Fig. 3 shows the number of enrolments and fails over the years; Phase 2 comprises the semesters 2015.2, 2016.1, 2016.2, and 2017.1. Compared with Phase 1, we can see a steady rise in the number of enrolments. During this phase, the increase in enrolments was followed by a higher number of fails, mostly due to students dropping out the module (i.e., not attending the minimum required number of lectures). In 2016.2, we had a peak on the number of fails (11 students). We do not have data to justify such a peculiar situation. The figures regarding 2017.1 are now shown in Fig. 3, since we have incomplete information about this semester due to the transition to teaching at a different university. Therefore, the conclusion of this module was carried out by another lecturer.

Theories, Tools, Materials, and Communities. Motivated by the limitations incurred by the use of CZT and Z-Eves (particularly, the ones associated with simulating and verifying specifications automatically), we decided to change from Z to Event-B. The latter is supported by the Rodin platform[10].

In addition to the standard features of an IDE (e.g., parsing, type checking, syntax highlighting, etc.), Rodin can be augmented by a number of plug-ins[11]. For instance, it is possible to integrate with ProB[12] [9] in order to animate and model check Event-B models. Moreover, graphical visualisation of formal models can be constructed using another plug-in: BMotion Studio (now, BMotion-Web[13]). It is important to mention that ProB is also capable of animating and verifying Z models. However, other interesting possibilities enabled by the Rodin platform and associated plug-ins are not available for Z; particularly, graphical visualisation of models via BMotionWeb. Although Rodin also has an interactive proof system, there was not enough time during the module to explore this feature. The verification of properties was carried out automatically via model checking (using ProB).

Concerning documentation, although the online community is limited to some extent (i.e., a relatively small number of projects and people using Event-B), there are an interesting number of tutorials, introductory materials and examples available online. The feedback collected from the students clearly indicates that they really appreciated this change.

During Phase 2, we also changed from teaching Formal Methods with CSP# to CSP_M. Although we prefer the combination of CSP with a functional language, we first opted for CSP# due to the limitations of FDR2, as explained before. With the advent of FDR3[14] [7] (now, FDR4) – a modern refinement

---

[10] Link: https://www.event-b.org/.

[11] Link: https://wiki.event-b.org/index.php/Rodin_Plug-ins.

[12] Link: https://prob.hhu.de/w/index.php?title=Main_Page.

[13] Link: https://prob.hhu.de/w/index.php?title=BMotion_Studio.

[14] Link: https://cocotec.io/fdr/.

checker for CSP_M, we revised our initial choice. The new version was also accompanied by a more comprehensive documentation, which is available online.

Teaching and Assessment Dynamics. At Phase 2, the teaching and assessment dynamics remained the same as Phase 1.

Impressions. We summarise our impressions from this phase as follows.

– We received a number of positive feedback concerning the adoption of Event-B and Rodin. Now, the students were able to specify, validate via animation and visualisation, and verify properties of formal models using a single formal notation.
– The integration of different formal notations (i.e., teaching Event-B and CSP in a single module) was still a challenge. As one student noted: when they were getting used with the first notation (Event-B) and tool support (Rodin), they needed to start from scratch and learn a different notation (CSP_M) and tool (FDR3).
– Regarding the seminars, in most semesters, the presentations delivered by the students were still shallow, as we highlighted in our impressions taken from Phase 1.

3.3   Phase 3: Teaching with B

After moving to a different university in 2017, the experience of teaching Formal Methods has been carried out at *Universidade Federal de Pernambuco* (UFPE[15]) to present days. UFPE is a federal public institution of higher education.

General Information. Both in the current (admissions starting from 2024) and in the older curricula, Formal Methods is an optional module. Software Engineering is a direct prerequisite. Additionally, Discrete Mathematics and Formal Logic are also indirectly required and, thus, the students cannot enrol at Formal Methods before the 5th semester. Nevertheless, a revision of these prerequisites is delivered to ensure a minimum knowledge of them. In the older curriculum, the name of the module is Critical Systems, whereas Formal Methods is adopted in the new one.

The main goal of the module is to discuss formal development of software and systems. This should be accomplished by illustrating practical applications of Formal Methods. The students should learn notions of formal specifications, refinement, and formal verification. Previously, this was a 75 h module; now, it comprises 60 h. Regarding capabilities that should be developed by the students, in essence, they are the same as described before for Phases 1 and 2.

Fig. 3 shows the number of enrolments and fails over the years; Phase 3 comprises the semesters ranging from 2017.2 to the present day. Differently from

---

[15] Link: https://www.ufpe.br/.

UPE, where the module was offered every semester, it is offered once a year at UFPE. The semester 2017.2 was an exception related to the transition from UPE to UFPE. The semester 2024.1 has not finished yet and, thus, we do not present data about it. Nevertheless, so far, we are observing figures that are consistent with the historical ones.

Compared with Phases 1 and 2, apart from 2019.1, we have a significantly lower number of enrolments. This is not surprising since this module is optional at UFPE. In 2017.2, we relate the slightly higher number of enrolled students with the attraction created by the module being offered for the first time by a different lecturer, adopting new teaching and assessment dynamics. In 2020.1, due to the COVID-19 pandemic, the module was delivered remotely in its entirety. This may have motivated more students to enrol at it. In 2021.1, we followed a blended teaching approach. Assessments were held exclusively in person. In the following semesters, all activities are delivered in person, as before the pandemic. This is mandatory according to the pertinent regulations. Concerning 2019.1, we cannot relate the peak of enrolments to any specific reasons.

Generally speaking, at Phase 3, the fails are mainly related to not attending the minimum required number of lectures. Our reading is that some students enrol at the module without a prior knowledge of its syllabus. During the first lectures, after understanding the associated expectations, they drop out of the module. The students that engage in the module's activities achieve pass marks.

Theories, Tools, Materials, and Communities. Motivated by the complaints about the loose integration between Event-B and CSP_M, and the time required to teach and learn the basics of both concepts, we decided to adopt a single notation for the whole module: the B language. Moreover, industrial application of B in the railway domain attracts the attention of students, promoting engagement.

Atelier B[16] is a tool that enables the operational use of the B method. It has a GUI equipped with the typical features of modern IDEs. Using commands (i.e., command mode) is also possible, which enables automated integration with other tools and methods. Nevertheless, this is not exploited in this module. Proof obligations are automatically generated and possibly discharged by the tool. Integration with external solvers is possible. Interactive proof development is supported by AtelierB. Nevertheless, so far, we have not had the time to explore this feature in this module. Nowadays, there is another module with an exclusive focus on proof development using the Coq [4] proof assistant. A nice feature of AtelierB is a visual correlation between fragments of the formal specification and proof obligations: one can clearly see the obligations associated with a given part of the formal model.

In Atelier B, abstract specifications can be refined into concrete implementations, from which C code is automatically generated. Although animation and visualisation are not supported by this tool, these features can be exploited by loading B specifications into ProB and BMotionWeb, respectively. Now, using a

---

[16] Link: https://www.atelierb.eu/en/atelier-b-tools/.

single notation (B), but also no longer addressing concurrency, and a combination of tools (Atelier B, ProB, and BMotionWeb), we teach our students about formal specification, animation, visualisation, verification, and code generation; thus, revisiting with rigour a broader perspective software engineering.

Although Event-B is also supported by Atelier B, we decided to focus on B due to other aspects, such as learning materials. In 2017, a Massive Open Online Course (MOOC) on B was made freely available online[17]. Other available materials are discussed in [8]. Unfortunately, concerning online communities, similar to Event-B, there are a relatively small number of projects and people using B. For example, compared with Coq, which seems to have a bigger community, it is not as easy to find online solutions to basic problems faced by the students.

Teaching and Assessment Dynamics. To increase the practical aspect of the module, now, all lectures are delivered at the laboratory. Generally speaking, the teaching dynamics is as follows. Before the class, the students should watch the associated video from the MOOC, prepare a mind map summarising the main concepts, and take notes of their doubts. During the class, we address the raised doubts and, whenever possible, illustrate how they relate to the use of the B supporting tools. We no longer have seminars and, thus, we would say that our teaching approach changed to a unified overview of Formal Methods, following the categorisation introduced in Sect. 1.

The marks are determined by composing the assessment of a project and an exam. The project has three deliverables, covering the possibilities of Formal Methods summarised in Fig. 2. The first deliverable is an abstract and formal model (i.e., a collection of Abstract Machines) created from an informal specification of a given problem. This model should have been properly animated and verified. The second deliverable encompasses the development of a graphical visualisation using BMotionWeb. Finally, after being confident that the right model has been built, in the last deliverable, the students are supposed to refine their Abstract Machines into Implementations, from which C code should be generated and executed.

During some semesters, we let the students propose the scope of their projects. However, this incurred some problems. Due to the lack of experience with Formal Methods, some of them went for a too ambitious scope. Now, the scope of the project is delimited by the lecturer. The projects are developed by groups of students. Differently, the exam is an individual assessment carried out at the laboratory using the B supporting tools.

Impressions. We summarise our impressions from this phase as follows.

– The students enjoy learning Formal Methods using B. As a language based on set theory and predicate logic, some students say that it demonstrates a practical need for teaching and learning Discrete Mathematics and Formal

---

[17] Link: https://mooc.imd.ufrn.br/course.

Logic. Nevertheless, there are some complaints about nuances of the syntax of B. It is also important to bear in mind that, since this module is optional, the enrolled students are naturally more inclined to learn Formal Methods than in a compulsory context.

- Additionally, going from abstract specifications to code within the same formalism is a distinguishing characteristic. Moreover, this is done by exercising different Formal Methods techniques.
- The difficulty of finding (online) answers to basic problems faced by the students is still a complaint.

4   Summary of Reflections and Recommendations

Table 1 summarises our experience on teaching Formal Methods for 10 years (i.e., during Phases 1–3) in Brazil. Regarding the impressions presented in the last row, the symbol - denotes negative impressions, whereas + indicates positive ones.

In Phases 1 and 2, the module was part of the Computer Engineering course. In Phase 3, the module was attended by students from different courses: Computer Engineering, Computer Science, and Information Systems. In all phases, the focus was on undergraduate studies. In terms of direct prerequisites, we had Software Analysis and Design (Phases 1–2) and Software Engineering (Phase 3). At UFPE (Phase 3), Discrete Mathematics and Formal Logic were also indirect prerequisites. At UPE (Phases 1–2), the enrolled students typically had already attended these two modules, even though this was not formally required.

Regarding the average (μ) number of students enrolled per semester, we can clearly see a reduction in Phase 3. This is expected, since, in Phases 1 and 2, the Formal Methods module was a mandatory one. The figures in Phase 1 are lower than the ones in Phase 2. Our hypothesis is that the module, as taught initially, was perceived as too difficult by the students, who decided to postpone it, since it was not a prerequisite to any other mandatory module. Considering the total ($\sum$) number of enrolments per phase, the fail ratio due to not reaching the minimum required mark is significantly lower in Phase 3 than in Phases 1 and 2. This is also expected. By being optional, the module should attract students interested in the area and, thus, willing to make the necessary effort to achieve pass marks. The fail ratio due to attendance issues is relatively high in Phase 3. Our hypothesis is that some students enrol at the module without prior knowledge of its syllabus. Then, they drop out of the module during the semester.

Teaching occurred mostly in person. Due to COVID-19 pandemic, during two semesters, it was necessary to first follow a remote approach, which then changed to a blended one. Our teaching and assessment dynamics are now oriented to deliver a very practical module based on flipped classrooms. Group projects have been developed inspired by toy examples (e.g., revisiting the classical lift example), research papers (e.g., an Air Traffic Control system), and simplifications of real systems (e.g., a High Frequency Jet Ventilator, an Automated External

Table 1. Summary of Phases 1–3.

	Phase 1	Phase 2	Phase 3
Period	2014–2015	2015–2017	2017–now
Institution	UPE (BR)	UPE (BR)	UFPE (BR)
Course	Comp. Engineering	Comp. Engineering	Comp. Engineering Comp. Science Information Systems
Level	Undergraduate	Undergraduate	Undergraduate
Module's name	Formal Methods	Formal Methods	Critical Systems
Curriculum	Mandatory	Mandatory	Optional
Prerequisites	SW Analysis and Design	SW Analysis and Design	SW Engineering
	Phase 1	**Phase 2**	**Phase 3**
μ enrolments	11 per semester	25 per semester	7 per semester
$\sum$ enrolments	22	77	52
Fails (mark)	18.18%	16.88%	5.77%
Fails (attend.)	4.55%	19.48%	19.23%
	Phase 1	**Phase 2**	**Phase 3**
Delivery mode	In person	In person	In person (mostly)
Methodology	Traditional	Traditional	Flipped classroom
Teaching	Classroom Laboratory sessions Seminars	Classroom Laboratory sessions Seminars	Laboratory sessions
Assessment	Project (2 parts) Written exams (2)	Project (2 parts) Written exams (2)	Project (3 parts) Exam at the lab.
Marking	Manual	Manual	Manual
	Phase 1	**Phase 2**	**Phase 3**
Approach	Fragmented overview	Fragmented overview	Unified overview
Theories	Z CSP#	Event-B CSP_M	B
Tools	CZT Z-Eves PAT	Rodin ProB BMotion Studio FDR3	Atelier B ProB BMotionWeb
Materials	Book [15] Tools doc.	Tools doc. Book [10]	MOOC of B Tools doc.
Communities	Scarce	Scarce	Scarce
Impressions	- Loose connection - Tools (Z) - Scarce community - Shallow seminars	- Loose connection + Tools (Event-B) - Scarce community - Shallow seminars	+ Various integrated applications of FMs + Tools (B) - Scarce community

Defibrillator, among others). Students' assessments are marked manually. The associated effort is relatively low, since we do not have large classes.

Over the years, we moved from a fragmented overview of Formal Methods to a unified one. This was motivated by the challenges to teach the integration of different formal methods to undergraduate students. The adopted languages and theories also changed during this period with the goal of presenting the students to user-friendly tools that also exercise various integrated applications of Formal Methods (i.e., specification, animation, verification, translation, and visualisation). The relatively scarce online community still imposes learning challenges.

We build on our experience and reflections to highlight aspects that should be taken into account when planning the teaching of Formal Methods as part of Computer Science education.

- **Curriculum**: thinking about correctness is crucial in modern software-based society. Since from the first semesters, we should discuss how basic modules (e.g. Discrete Mathematics and Formal Logic) contribute to the design of safe and secure systems. Furthermore, considering its important role, teaching Formal Methods should be an integral part of Computer Science education.
- **Module scope**: the name of the module should be Formal Methods to give visibility and empower the area. If possible, this should be a mandatory module to be taught after Software Engineering. We recommend adopting a unified overview approach, which could be followed by other optional modules that focus on specific techniques (e.g., model checking, theorem proving, code analysis) and trends (e.g., the application of formal methods to intelligent systems). A fragmented overview of Formal Methods may be interesting to graduate students, which may already have some background in the area.
- **Theories, tools, materials, and communities**: the introductory module on Formal Methods should be more practical than theoretical. The details of the underlying theories should be addressed in subsequent modules. Appropriate languages and tools should be chosen. Tools should be user-friendly. If not readily available, learning materials (e.g., multiple examples, frequent questions and answers, frequent errors and solutions) should be prepared in advance, since the information available online may be limited.

5   Conclusion and Perspectives

In this paper, we build on our 10-years experience to discuss possibilities to the teaching of Formal Methods. In the following sections, we briefly comment on related works, and provide our final considerations.

5.1   Related Work

This paper contributes to the thoughts on how to teach Formal Methods, many of them shared in the proceedings of the Formal Methods Teaching workshop[18].

---

[18] Link: https://dblp.org/db/conf/tfm/index.html.

We draw the reader's attention to three recent publications on teaching with Event-B [5,6] and B [8]. In [6], the authors provide arguments for following the opposite direction as considered here; that is, moving from B to Event-B. Aligned with our impressions, these previous experiences also emphasise the importance of tools and proper education in Formal Methods to the development of correct implementations. Delivering practical and project-oriented teaching is also mentioned as a key point.

5.2 Perspectives

Besides continue monitoring the teaching of Formal Methods at CIn-UFPE, we plan to: (i) articulate with the lecturers of other modules (e.g., Discrete Mathematics, Formal Logic, and Software Engineering) to strengthen the importance of learning Formal Methods, (ii) enrich our learning materials in order to anticipate the problems faced by the students, and (iii) collect data on how studying Formal Methods contributed to students' careers.

Acknowledgments. This work is partially supported by INES (www.ines.org.br), CNPq grant 465614/2014-0, CAPES grant 88887.136410/2017-00, and FACEPE grants APQ-0399-1.03/17 and PRONEX APQ/0388-1.03/14.

Disclosure of Interests. The author has no competing interests to declare that are relevant to the content of this article.

References

1. Abrial, J.R.: The B-Book: Assigning Programs to Meanings. Cambridge University Press (1996)
2. Abrial, J.R., Butler, M., Hallerstede, S., Hoang, T.S., Mehta, F., Voisin, L.: Rodin: an open toolset for modelling and reasoning in Event-B. STTT **12**(6), 447–466 (2010). https://doi.org/10.1007/s10009-010-0145-y
3. ter Beek, M., Bory, M., Dongol, B., Sekerinski, E.: The role of formal methods in computer science education. Tech. rep. (2023). https://csed.acm.org/wp-content/uploads/2023/11/Formal-Methods-Nov-2023-1.pdf
4. Bertot, Y., Castran, P.: Interactive Theorem Proving and Program Development: Coq'Art The Calculus of Inductive Constructions, 1st edn. Springer Publishing Company, Incorporated (2010)
5. Cataño, N.: Teaching formal methods: lessons learnt from using Event-B. In: Dongol, B., Petre, L., Smith, G. (eds.) FMTea 2019. LNCS, vol. 11758, pp. 212–227. Springer, Cham (2019). https://doi.org/10.1007/978-3-030-32441-4_14
6. Dubois, C., Prevosto, V., Burel, G.: Teaching formal methods to future engineers. In: Dongol, B., Petre, L., Smith, G. (eds.) FMTea 2019. LNCS, vol. 11758, pp. 69–80. Springer, Cham (2019). https://doi.org/10.1007/978-3-030-32441-4_5
7. Gibson-Robinson, T., Armstrong, P., Boulgakov, A., Roscoe, A.W.: FDR3 — A modern refinement checker for CSP. In: Ábrahám, E., Havelund, K. (eds.) TACAS 2014. LNCS, vol. 8413, pp. 187–201. Springer, Heidelberg (2014). https://doi.org/10.1007/978-3-642-54862-8_13

8. Lecomte, T.: Teaching and training in formalisation with B. In: Dubois, C., San Pietro, P. (eds.) Formal Methods Teaching, pp. 82–95. Springer Nature Switzerland, Cham (2023). https://doi.org/10.1007/978-3-031-27534-0_6

9. Leuschel, M., Butler, M.J.: PROB: an automated analysis toolset for the B method. STTT **10**(2), 185–203 (2008). https://doi.org/10.1007/s10009-007-0063-9

10. Roscoe, A.W.: Understanding Concurrent Systems. Texts in Computer Science, Springer (2010). https://doi.org/10.1007/978-1-84882-258-0

11. Shi, L., Liu, Y., Sun, J., Dong, J.S., Carvalho, G.: An analytical and experimental comparison of CSP extensions and tools. In: Aoki, T., Taguchi, K. (eds.) ICFEM 2012. LNCS, vol. 7635, pp. 381–397. Springer, Heidelberg (2012). https://doi.org/10.1007/978-3-642-34281-3_27

12. Sun, J., Liu, Y., Dong, J.S., Pang, J.: PAT: towards flexible verification under fairness. In: Bouajjani, A., Maler, O. (eds.) CAV 2009. LNCS, vol. 5643, pp. 709–714. Springer, Heidelberg (2009). https://doi.org/10.1007/978-3-642-02658-4_59

13. The White House: back to the building blocks: a path toward secure and measurable software. Tech. rep. (2024). https://www.whitehouse.gov/wp-content/uploads/2024/02/Final-ONCD-Technical-Report.pdf

14. Woodcock, J., Cavalcanti, A.: The semantics of *Circus*. In: Bert, D., Bowen, J.P., Henson, M.C., Robinson, K. (eds.) ZB 2002. LNCS, vol. 2272, pp. 184–203. Springer, Heidelberg (2002). https://doi.org/10.1007/3-540-45648-1_10

15. Woodcock, J., Davies, J.: Using Z (1996). http://www.usingz.com/

An Educational Module for Temporal Features in Alloy 6

Luca Padalino⬤, Francesca Pia Panaccione⬤, Francesco Santambrogio⁽✉⁾⬤,
Elisabetta Di Nitto⬤, and Matteo Rossi⬤

Politecnico di Milano, Milano, Italy
{luca.padalino,fracescapia.panaccione,francesco2.
santambrogio}@mail.polimi.it, {elisabetta.dinitto,matteo.rossi}@polimi.it

Abstract. As software systems have become increasingly important, teaching Software Engineering students how to develop high-quality software is essential. In this regard, formal modeling and verification are important educational tools that help students in getting an in-depth understanding of software. Nonetheless, formal languages are not straightforward to teach and, therefore, carefully designed materials are needed to convey them. In this paper we focus on Alloy, which is an easy-to-learn formal language equipped with a usable analyser, and we present a complete teaching module that can be used by teachers to support students in learning the temporal constructs defined in its latest version, Alloy 6. The module is designed exploiting active learning methods and is supported by multimedia content. It is openly available and can be reused and tailored to the need of specific courses.

Keywords: Alloy 6 · formal methods · modeling language · active learning · teaching module

1 Introduction

With the increasing importance of software systems in our daily lives, Software Engineering students must be provided with the knowledge and skills they need to develop high-quality software, which increasingly relies on formal methods for verifying design, safety, and functionality. In particular, formal specification languages are essential for accurate modeling and verification and, thus, to guide the definition of requirements and to facilitate smooth design, implementation, and testing processes.

Among formal specification languages, Alloy [9] stands out as having one of the simplest syntaxes to read and write, yet with considerable expressive power [15]. In Alloy, a system can be represented using a collection of types defined through *signatures*, each having different fields, governed by rules and constraints defined as *facts*. Alloy has been widely applied in the literature,[1]

---

[1] alloytools.org/applications.html.

© The Author(s), under exclusive license to Springer Nature Switzerland AG 2024
E. Sekerinski and L. Ribeiro (Eds.): FMTea 2024, LNCS 14939, pp. 75–90, 2024.
https://doi.org/10.1007/978-3-031-71379-8_5

and it is supported by a mature tool, the Alloy Analyzer, which enables fully automated system analysis and can reveal weaknesses early on and promote incremental development. These features make Alloy an appealing language to introduce students to the problem of formally specifying real-life systems. Alloy has recently introduced the sixth version[2] where linear temporal logic constructs have been added to the language. This has enriched Alloy's expressive power, but, at the same time, has introduced new teaching and learning challenges.

The contribution of this paper is the development of a teaching module designed to present the temporal features of Alloy 6 to students who have been already introduced to the basic features of Alloy. The module includes slides, Alloy models, videos, and guidelines for instructors. The material is meant to be used within a traditional introductory lecture, a flipped class, an exercise session, and a challenge to encourage students to explore the practical application of the approach. Moreover, the material is organized in several parts that can be used and—possibly—tailored independently from the others. The material is openly available on our Github repository.[3] So far, it has been used, in a shortened format, in the Software Engineering 2 course at Politecnico di Milano in the Academic Year 2023–2024. As a preliminary evaluation, we report on this experience commenting on the performance of students during the exam both in terms of achieved score and in terms of types of errors made. This analysis shows that the module is potentially effective.

The paper is structured as follows: Sect. 2 presents the state of the art on the preparation of teaching modules and on Alloy. Section 3 presents an overview of the temporal features of Alloy 6. Section 4 presents the proposed teaching module and Sect. 5 evaluates it. Finally, Sect. 6 presents a critical discussion and concludes the paper.

2   State of the Art

Teaching Modules. A teaching module is a significant, highly homogeneous and unified part of a planned disciplinary program. One of the main characteristics of a teaching module is having well-defined and verifiable learning objectives which not only enhance student engagement and motivation [18], but also assist instructors in designing assessments and selecting the content of the lectures effectively.[4] To achieve such goals, educators must consider how to help students take advantage of the contents by identifying effective teaching materials and strategies.

With the advent of educational technology, universities are increasingly integrating instructional multimedia into course delivery as part of the teaching materials, such as slides, videos, or online quizzes, providing several advantages including increased access to content, personalized learning opportunities, and greater student engagement [19]. Multimedia-enhanced learning environments

---

[2] alloytools.org/alloy6.html.

[3] github.com/lucapada/ResearchProjectAlloy6.

[4] cteresources.bc.edu/documentation/learning-objectives.

can foster student motivation and facilitate problem-solving skills through self-exploration and collaboration [12]. In the context of teaching modules, teaching strategies aim to overcome the limitations of traditional classroom teaching, where students often passively consume information, and leverage active learning instead, which fosters students' engagement and encourages them to take responsibility for their learning [11].

Howell [8] discussed student experiences and perceptions of an interdisciplinary social science course, concluding that over 90% of respondents agreed that in-class active learning exercises made the classes more engaging and the material more memorable than usual. Active learning includes projects, problem-solving tasks, and team assignments; it offers numerous benefits, such as receiving immediate feedback, building confidence, and promoting cognitive development [7]. Accordingly, it is suitable for disciplines with practical aspects like software engineering and specification languages in particular. The rest of this section reviews some of the most useful active learning strategies explored and tested in the literature.

One of the most used and simplest active learning strategies is questioning, which consists in having teachers pose proper questions to learners. They encourage discussion, argumentation, and the expression of opinions and alternative views. When used effectively, questioning provides immediate feedback about students' understanding, supports informal assessment, and evaluates teaching strategies' impact [5]. Questioning can be seen as a form of feedback for the students who can measure their knowledge against the asked questions and required answers. In general, feedback is an active learning strategy that informs a student or a teacher about their performance and is acknowledged as an essential element for improving the students' learning process [6]. Feedback redirects or refocuses students' actions so that they can align effort and activity toward a clear outcome [5].

Besides teaching theoretical aspects and involving students with questioning and feedback, examples may help them understand better the concepts of the lectures. Worked examples, in particular, aid initial cognitive skill acquisition by presenting formulated problems, solution steps, and final solutions [17]. Studying worked examples is effective for teaching complex problem-solving skills as it provides expert mental models to novices and reduces cognitive load, facilitating skill acquisition [20]. Once worked-examples are assimilated, students can push toward more complex real-world problems. Problem-based learning is the instructional approach where learning occurs through the process of solving such problems [2]. Allen et al. concluded that such a method enhances the affective domain of student learning, improves student performance on complex tasks, and fosters better retention of knowledge [1]. This strategy exploits projects that should be as complete as real-world instances so that students can experience the whole process that is likely to be seen in future working environments, including cooperation with other peers.

In general, as stated by Laal and Ghodsi [10], there are several benefits brought by collaborative learning, like enhancing interactions, learner autonomy, teamwork, and problem-solving skills within a group, as members exchange ideas and collectively build shared understanding. Integrating communication, interaction, and cooperation skills among team members is crucial for successful software development. This innovative approach prepares students for future industrial settings by emphasizing the significance of social aspects in software development.

In the development of our teaching module we have used and adapted to the specific context of Alloy learning all the strategies presented in this section. We discuss the implementation of these features in Sect. 4.

Teaching Material for Alloy 6. Alloy has been extensively integrated into undergraduate and graduate courses worldwide. Examples of courses adopting it can be found in the Formal Methods Europe (FME) database[5] of formal methods courses. From an analysis of the syllabus and, where possible, the material, it results that most of the listed courses do not focus on the temporal features offered by Alloy 6. Notable exceptions are the Formal Methods for Software Engineering course,[6] taught at University of Minho (Portugal) by the group that contributed to the development of Alloy 6 (and the Alloy4Fun web application [13]), the Logic for Systems course at the Brown University,[7] and the University of Iowa's Formal Methods in Software Engineering course.[8].

In all the cases we could analyse, the offered teaching material was not fully self-contained and organized to be reused in other classes. Also, the presented examples were limited in number and relatively simple. As such, we concluded that a comprehensive teaching module addressing the complexities of Alloy's temporal features can be useful to aid Software Engineering teachers in incorporating such features in their courses and to help students in mastering their application in significant cases.

3   Overview of Alloy's Temporal Features

Alloy 6 introduces the concept of *mutable* signatures and fields, enabling users to express their temporal evolution using operators derived from Linear Temporal Logic (LTL, [16]). In a nutshell, a mutable signature captures a set whose members can change over (discrete) time; similarly, if a signature (not necessarily a mutable one) includes a mutable field, then an instance of the signature could be such that the value of the field changes over time. Mutable signatures and fields are identified through the `var` keyword. Alloy uses a discrete notion of time, so for each time instant $i \in \mathbb{N}$ the instance of a mutable element can be different (signatures and fields that are not marked as `var`, instead, have the same value for each time instant).

---

[5] fme-teaching.github.io/courses.
[6] haslab.github.io/MFES.
[7] csci1710.github.io/2024.
[8] homepage.cs.uiowa.edu/~tinelli/classes/181/Fall23/index.shtml.

Alloy includes the classic LTL temporal operators (including their past counterparts), such as `always`, `eventually`, `until`. The LTL "next" operator is called `after` in Alloy, and `before` is its past counterpart. For example, the following declarations define that signature S is mutable, initially it does not contain any elements, but at some point it will contain 3.

```
var sig S{}
fact { #S = 0 and eventually #S = 3 }
```

Alloy 6 can perform both *bounded* and *unbounded* model checking of temporal specifications (the latter through the nuXmv formal verification tool [4]). If a bounded approach is used [3], the time horizon (i.e., the maximum length of the traces to be explored in search for a loop) must be provided through the `steps` keyword.

Alloy 6 also features an improved visualizer tool which displays traces in a user-friendly way. More precisely, the visualization pane is split in two parts showing the model instance in consecutive states. The visualizer allows users to explore traces (move along the current trace, explore a new one, etc.) through suitable commands.

4   Organization of an Alloy 6 Teaching Module

The developed teaching module incorporates various teaching strategies (see Sect. 2) and focuses on the temporal features introduced in Alloy 6. It assumes that students are already familiar with the core notions of the Alloy modeling language (signatures, facts, predicates, etc.), but not with its temporal features. It includes theoretical and exercise lectures, within which a timeline was defined for each macro-topic to be presented. To experiment with alternatives to classic lectures, additional activities (self-assessment quizzes, a flipped classroom, a challenge) were added, which function as tools to assess comprehension related to learning objectives. Additionally, a comprehensive guide with an accompanying video was developed to assist instructors in understanding how to use the module effectively. The organization of the teaching module is presented in Fig. 1.

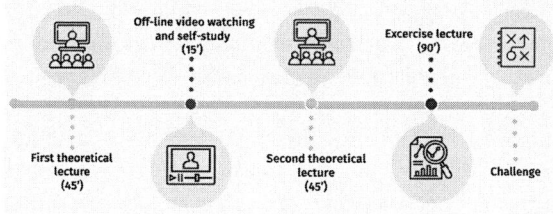

Fig. 1. Temporal organization of the teaching module.

First Theoretical Lecture. In the first theoretical lecture, the focus is on comparing how time-varying systems are handled in Alloy 5 vs. Alloy 6. The aim is to demonstrate the enhanced effectiveness of Alloy 6 in this regard, thanks to the introduction of new logic operators and keywords. By the end of this lesson, students should grasp how Alloy 5 addresses dynamic modeling, recognize its limitations and understand why the new features in Alloy 6 are necessary. The lecture begins with an overview of the learning objectives, setting the stage for what students will gain from the session. Then, the distinction between the static and dynamic world in Alloy is explored, providing clarity on the significance of these terms within the context of modeling. The bulk of the lecture focuses on examining how dynamic models are represented in Alloy 5, which implies the presentation of the Alloy ordering package and of the `Time` signature and its limitations. Through examples and explanations, students gain insights into the challenges faced when modeling dynamic systems using Alloy 5. Then, the focus shifts to Alloy 6, introducing the `var` keyword and LTL concepts and operators. These additions enhance Alloy's capabilities in handling temporal aspects and introduce mutability, addressing some of the shortcomings observed in Alloy 5. Throughout the lecture, the emphasis is on providing clear explanations and practical examples to aid understanding.

Video Watching and Self-studying. Video watching and self-studying help students learn the topics related to temporal operators introduced in Alloy 6. Students should have attended the first lecture, which introduces the features used in the video. The teacher, in the recorded video, unidirectionally explains the concepts, leaving the opportunity to raise doubts and gather opinions to the next lesson, through feedback and questioning. The recording focuses on temporal connectives—both future and past—that a user can exploit to define predicates, facts, and assertions. All connectives are presented in the same way, always following the same pattern: syntactic definition, semantics, and application example. As teaching material, the teacher provides the video and the slide set containing the whole content of the video.

Second Theoretical Lecture. In the second lesson, the objectives are two-fold: firstly, through the flipped class approach, to assess students' understanding of the material covered in the first lesson and in the video; secondly, to continue exploring the new temporal features introduced in Alloy. Students will have the opportunity to evaluate their comprehension and the effectiveness of their study methods through quizzes administered partly during class time and partly as take-home assignments. By the end of the class, students will gain a comprehensive understanding of Alloy's temporal features and capabilities. They will be able to use temporal connectives and address related arguments.

Exercise Lecture. During the exercise lecture, students will engage in hands-on activities designed to reinforce their understanding of Alloy's temporal concepts.

Through a worked-example approach, the lecturer will lead students through practical coding exercises aimed at tangibly applying theoretical knowledge. This interactive session will provide a valuable opportunity for students to deepen their comprehension and develop essential problem-solving skills. To complete the exercises, students must have attended the lectures covering Alloy's new features, watched the related video on connectives, and installed the latest version of the Alloy tool. The exercise session comprises the following three exercises.

Exercise 1: Concurrent Communications in Distributed Systems This exercise delves into the concept of parallel system operations, focusing on modeling communication in distributed systems. Students will have to define mutable signatures and establish facts defining the conditions that are perpetually true as well as the ones that describe the way the system evolves over time. By modeling scenarios such as message transmission and reception, students will gain insight into handling dynamic systems within the Alloy framework.

Exercise 2: Travel (Interrail) In this exercise, students will tackle the modeling of travel scenarios using Alloy. They will define signatures representing entities involved in travel, such as cities and travelers, and create static and dynamic models of travel itineraries. By modeling a person's journey between cities and defining completion criteria, students will gain proficiency in modeling dynamic real-world processes.

Exercise 3: Mailbox The final exercise focuses on modeling a mailbox system with dynamic behavior. Students will define a model for messages that can be deleted and restored from a recycle bin, with their locations changing over time.

Throughout the exercise session, students will actively engage with the material, applying theoretical concepts to practical scenarios. By completing these exercises, students will enhance their understanding of Alloy's capabilities and develop the skills necessary for effective modeling and analysis.

Challenge. The extra activity offered by the module is a challenge where students can apply what they have learned in solving a particular specification within a two-week time frame. To tackle the challenge and solve the exercise, students must have previously (i) attended the previous lectures, (ii) watched the video related to connectives, which are extensively used when defining predicates, facts, and assertions, (iii) attended the exercise session and (iv) installed the latest version of the Alloy tool. The challenge presents students with a real and current problem that may not have a single solution. Problem-solving, a skill greatly developed by the challenge, encourages each individual to propose a personal solution based on what they have learned. In this proposed challenge, the strategy of collaborative learning is employed, with students working together in small groups (two or three students) to develop skills related to teamwork and problem-solving. The proposed theme for the challenge is Software-Defined Networks, inspired by [14], which are increasingly important in cloud computing and network architectures to facilitate their administration, configuration and monitoring, and to improve their performance.

For what concerns the teaching material, the teacher provides the slide set containing the outline of the challenge and how to deliver it. At the end of the challenge—thus, at the end of the proposed module—students will be able to face situations in which they are required to model complex systems by working in a group in a very practical and professional setting.

Each component of the teaching module was carefully selected to maximize its effectiveness, through the active learning strategies described in Sect. 2. Table 1 outlines the various components of the module, detailing the specific teaching strategies and goals associated with each part.

Table 1. The various components of a teaching module and the associated teaching strategies and goals within the module.

Module's component	Teaching strategy
First theoretical lecture Second theoretical lecture	Feedback Questioning Learning goals setting
Video watching and Self-study	Multimedia teaching material usage
Exercise lecture	Questioning Feedback Worked example Problem-based learning
Challenge	Problem-based learning Collaborative learning

5   Evaluation

In this section, we present a preliminary evaluation based on the module's adaptation and usage within the Software Engineering 2 (SE2) course at Politecnico di Milano. The course is mandatory for students enrolled in the Master in Computer Science and Engineering. It is offered during the Fall semester (September-December) to more than 600 students divided in three classes. Most of the students enrolled in the SE2 course have a background (typically a Bachelor's degree) in computer science and engineering, though there are also students of other engineering disciplines such as Telecommunications Engineering and even Mechanical Engineering.

Alloy was presented as a specification language for model definition and verification at the requirement analysis and design levels, and its sixth version was introduced in the course during the 2023–2024 academic year. Due to time constraints, the module was condensed and adapted to fit in a 1-hour lecture, plus a half-hour exercise session. Hence, a full comparison of the differences between the new temporal features and the approaches previously available in Alloy (see Sect. 3) was not possible. The condensed lecture, instead, focused on how the new temporal features of Alloy can capture, in a natural way, the change in the state of a system (i.e., in its mutable parts) when certain operations are performed. In addition, it provided, through examples, an overview of the most commonly used temporal operators (`after`, `aways`, etc.), and showed how they can be used to describe constraints on the evolution of mutable features.

The exercise session provided further examples of definitions of temporal properties (including assertions to be checked) through the development of a complete, albeit small, specification of a message handling system, a variation of the mailbox exercise of the teaching module (see Sect. 4). In a way, the exercise session replaced the challenge, which was created to drive students to solve (possibly through group work) a particularly long and complex problem. The introduction of this new material did not result in an increase of the number of hours dedicated to Alloy in the course. To make optimal use of the available class time, we showed how the new temporal features in Alloy 6 allowed for a clearer description of the relations between states before and after the execution of operations with respect to Alloy 5 (where such relations are described through the introduction of specific atoms capturing different instances of the state).

In the context of the SE2 course, the negative impact, teaching-wise, of dropping the challenge was lessened by how assessments are carried out. A significant portion of the students enrolled in the course elect to acquire part of the course credits through the development of a project focusing on the requirements analysis and design activities related to a given application. As part of these activities, students are asked to create a working Alloy specification, which is another form of challenge. Finally, the vast majority of students enrolled in the SE2 course complete the course credits by taking a written exam, which includes an Alloy-specific question requiring students to define a small number of signatures and properties.

To evaluate the effectiveness of teaching the temporal aspects of Alloy using the material derived from our module, we have analyzed only the performance of students during the written exam. This choice was driven by the fact that written exams are individual, thus more objective and more comparable across different course editions. We performed two types of analyses. First, we collected the scores achieved by students in the Alloy question and compared them with those obtained in the previous 4 years (in which temporal features were not presented). Then, we selected a subset of 50 exams from this year and we manually examined them to identify the most common errors. This allowed us to highlight the aspects where students are weaker and require more support. For each academic year, scores were manually assigned by the same two course instructors (who each has been teaching the SE2 course for the past 7 years). Each instructor graded roughly half of the exams. The analysis carried out for this work does not include feedback provided by students, which we plan to systematically collect in future editions of the SE2 course.

Analysis of Students' Scores. We considered the scores of the Alloy question of the exams taken by students in the academic years from 2019–2020 to 2022–2023. We compared them with the scores of the exams taken in the current academic year, in which the Alloy question explicitly required to use the new temporal features. Our hypothesis is that the grade distribution is the same across years, even after introducing the new Alloy 6 features, thanks to the adoption and tailoring of the teaching module.

Table 2 provides an overview of the number of students considered in the analysis and the corresponding average scores. Notice that, at Politecnico di

Milano, for each course there are 5 exam calls spread across the academic year (2 in January-February, 2 in June-July, and 1 in September), and students can take the exam—possibly multiple times—in any of the 5 calls. The analysis was carried out focusing on 2 of the 3 classes (totalling more than 400 students) in which students are divided. For each considered academic year, only the February call was taken into account; for the SE2 course this is typically the second-most attended call (after the January one), since the course is taught in the Fall semester.

The first columns of Table 2 focus on the five academic years under consideration and show that the number of students taking the exam every year is statistically relevant and relatively stable between 125 and 153 and that the average score, computed in terms of percentage of correctness, is between 60% and 73%, with the lowest score obtained in the year 2020-2021. The sixth column provides the cumulative results obtained in the first four considered academic years and shows that the average score obtained this year (68%) is not significantly different from that of previous years (69%). In Fig. 2a scores are organized in the six categories defined in Fig. 2b. The distribution of scores in Fig. 2a highlights the bad performance obtained in the year 2020-2021, with 22% of low scores; also, it shows that this year's performance is in line with that of previous years, with most scores in the "medium" category.

Table 2. Number of students taking the exam (February call only) and average score expressed in terms of percentage of correctness.

	19–20	20–21	21–22	22–23	23–24	no time op.	time op.
#stud	130	143	153	125	142	551	142
Average	72%	60%	73%	72%	68%	69%	68%

In conclusion, our hypothesis that this year's grade distribution is similar to that of previous years can be considered confirmed, even though the observation of future academic years is needed to consolidate it.

Analysis of Students' Answers. We evaluated the exam held in February 2024. The text of the question had two parts, as shown below:

Consider a system to monitor the accesses of vehicles to the center of a city (think "Area C" in Milano). The system detects vehicles entering the City Center Area (CCA for short) using "gates" installed on the streets through which vehicles can access the CCA. Each time a vehicle goes through a gate, the gate reads the license plate of the vehicle and provides the system with the corresponding information (license plate number and time of passage of the vehicle). The CCA is active only until a certain time of the day (e.g., until 7pm). You are asked to use the features of Alloy 6 to capture some features of the system, focusing on the handling of the accesses for a single day. In particular, consider the (simplified)

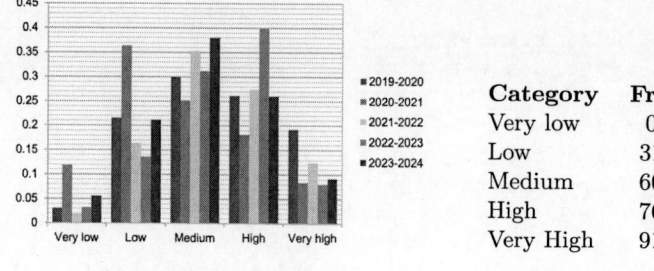

(a) Scores classification.

Category	From	To
Very low	0%	30%
Low	31%	59%
Medium	60%	75%
High	76%	90%
Very High	91%	100%

(b) Definition of categories.

Fig. 2. Scores per year, organized by categories.

domain model for the system shown in Figure 3 and represented through a UML Class Diagram.

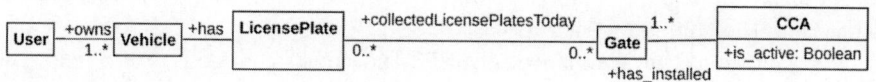

Fig. 3. UML class diagram provided in the exercise.

(Q1): *Define suitable signatures and constraints to capture the domain model shown above. In particular, identify the elements of the model that are mutable.*

(Q2): *Define a fact* activeCCAdef *that states that the CCA is initially active, then at some point it must become inactive; after becoming inactive, the CAA cannot become active again.*

The proposed solution to the two questions is shown in Listing 1.1. To assess the correctness of the solutions, the following elements were considered: the correspondence between the Alloy model and the UML class diagram, the appropriate identification of mutable elements, and the correct definition of the activeCCAdef fact.

Listing 1.1. Proposed solution to the exam questions.

```
sig User { owns : some Vehicle }
sig Vehicle {
    has : one LicensePlate
}{ one u : User | this in u.owns }
sig LicensePlate {}{ one v : Vehicle | v.has = this }
sig Gate {
    var collectedLicensePlatesDay : set LicensePlate
}{ one cca : CCA | this in cca.has_installed }
```

```
sig CCA {
    has_installed : some Gate
    var is_active : Boolean
}
fact activeCCAdef {
    all cca : CCA | cca.is_active = True
        and eventually cca.is_active = False
        and always (cca.is_active = False implies
            always cca.is_active = False)
}
```

We manually analyzed the answers produced by 50 students. This sample was selected from the batch of 142 exams graded by the course instructors, using the following procedure: (i) the exams of those students who obtained a score lower than or equal to 50% of the total were eliminated, as the nature of the errors in these cases certainly went beyond simply the usage of temporal operators; (ii) for each score between 50% (excluded) and 100% (with increments of 10%), a roughly equal number of exams per score was randomly extracted. The answers were finally made anonymous to allow for the sharing of the exams with people other than the course instructors.

We analyzed the exams to identify the most common errors and the corresponding topics in the teaching material, we investigated the reasons why students could have missed those concepts, and we checked whether the proposed teaching module offers material in that regard and could help. The analysis evidenced the following three main categories of errors. *Mutable signature* var. Fifty percent of the students (25 over 50) did not consider that the set of license plate numbers read when passing through the gate (attribute `collectedLicensePlatesDay` in Listing1.1) and the status of the CCA (attribute `is_active`) should be defined as mutable using the keyword `var`. This is an error that the majority of students made in only one of the two attributes to be defined as mutable. Moreover, this is the only kind of error made by many of the students who correctly defined the fact, thus suggesting that the main cause for the error was a lack of focus. However, this also highlighted that the teaching material should stress the importance of the `var` keyword, which is the starting point from time-dependent specifications. The number of examples and remarks made during theoretical lectures could be increased, especially when presenting temporal connectives. In this respect, the original module is more complete compared to the tailored one, as it presents multiple examples that focus, as a first step, on the importance of identifying what is mutable and what is not, and only after this define facts that use temporal connectives.

Time *signature definition.* Half of the students (25 over 50) created an ad-hoc Time signature as the basis for time-dependent properties. This is similar to how such properties are introduced in Alloy 5, a method made obsolete by the new features of Alloy 6. The following is an example found in one exam:

```
sig Time {hour: one Int}{ hour > 0 and hour < 24 }
sig Information {   plate: one LicensePlate ,
                    time: one Time   }
```

Most likely, students who followed this approach took the course in the previous academic year and did not take care to study the new aspects introduced this year. Indeed, the tailored material does not include a history of the evolution of the representation of time-dependent features from the previous versions of Alloy (5 and before) to the latest one; this is instead available in the complete module, which explains how defining a Time signature is no longer necessary in Alloy 6 thanks to the introduction of mutable features.

Usage of temporal connectives and definition of the fact. Seventy-eight percent of students (39 over 50) made conceptual, syntactical, and logical errors regarding the definition of facts. More than half used an incorrect temporal connective. Given the heterogeneity of errors, it was particularly complex to find a correlation among all the errors and a single reason. Not considering syntactic errors—which typically originate from a superficial study of the language syntax—we separated conceptual errors concerning the usage of first-order logic constructs from errors inherent to the usage of temporal connectives. For example, the errors in the following definition of fact activeCCAdef are mostly first-order logic-related and show that the student did not grasp the semantics of the all and some quantifiers. Having misunderstood these basic aspects, the student wrote a fact that is also incorrect from the temporal viewpoint.

```
fact activeCCAdef {
   all c: CCA | c.is_active
   some c: CCA | not c.is_active
   all c: CCA , g: Gate | not c.is_active implies
                     c'.has_installed = g and c'.is_active }
```

In the following example, the student used the temporal operators in a correct way, but missed the quantification on variable c:

```
fact activeCCAdef {
   always(c: CCA | c.is_active in True implies
                     eventually(c.is_active in False))
   and always (c: CCA | c.is_active in False implies
                     always(c.is_active in False)) }
```

Students might have misunderstood the meaning of each temporal connective and this may have impacted on the definition of the fact. In this regard, the module offers examples and quizzes for each temporal connective, with exercises focusing on why a certain connective is used and others should not. The examples presented in the theoretical lectures also help to better integrate the novelty of the temporal connectives with the existing logical constructs.

Threats to Validity. The experiments and the analysis presented in this paper are susceptible to the following threats to validity, which we plan to overcome in the future.

The evaluation of the teaching module from the instructors' perspective has been performed only internally, within the same group that has originated the module. A more in-depth evaluation involving instructors from multiple different academic institutions will be targeted for the next academic years.

The evaluation of the teaching module from the students' perspective is indirect, since the class that was the object of the evaluation of Sect. 5 was not directly exposed to the complete teaching module, but only to its tailoring (even though the whole module was available to them). This threat is mitigated by the fact that, since the complete module is more detailed (especially from the practical viewpoint) compared to the tailored one, we expect the students' learning experience to actually improve through exposure to the complete module.

Finally, the analysis has considered a group of students who attend the same study course, hence have a homogeneous background. It could be beneficial to carry out experiments with a more diverse student population, though this might be difficult to achieve.

6   Discussion and Conclusion

For software engineering students, understanding formal specification languages for requirements modeling is very important, though far from trivial. Alloy is an educationally accessible modeling language that has recently introduced important new features that revolve around an integrated concept of time and do not require the use of external modules; they include new keywords for marking objects and their properties as mutable, new operators for expressing properties related to past and future time instants, and a new visualizer tool.

We proposed a module to effectively teach students the new Alloy features by combining different teaching strategies, such as frontal lectures and flipped classrooms, to stimulate students' interest and help them better understand the language.

Additional research is needed to further validate the proposed approach. Experiments could be carried out to test the effectiveness of the teaching module objectively. In particular, we plan to compare different teaching methods by dividing a class into two groups: one group testing the module, while the other uses a different teaching approach.

Quizzes, exercises, and the challenge can be used to build an up-to-date picture of the state of student learning. Thanks to them, one could observe the improvement achieved by the class over time in terms of: (i) timeliness of students' responses, (ii) obtained scores, and (iii) quality of answers in relation to the used teaching method. In particular, the first two factors can be evaluated through quizzes and quick exercises, while the third is assessed through activities such as drills, challenges and written exams. The quality of answers, in particular, concerns how well students manage to tackle a complex problem from scratch, without teacher guidance.

As discussed in Sect. 5, to increase the objectivity of the analysis, we plan to carry out experiments in classes taught by different teachers and taken by students with different backgrounds.

We believe that our approach to teaching Alloy can be an important starting point for improving the accessibility of formal specification languages in software engineering; however, further research is needed to confirm the validity of our solution. Depending on the results of the validation, the proposed teaching module can be modified and fixed to be as effective as possible.

Acknowledgements. We are very grateful to all students involved in the SE2 course and to Alessandra Viale, who helped us with the preparation of the exams for the detailed analysis.

References

1. Allen, D.E., Donham, R.S., Bernhardt, S.A.: Problem-based learning. New Dir. Teach. Learn. **2011**(128), 21–29 (2011)
2. Barrows, H.: The essentials of problem-based learning. J. Dent. Educ. **62**(9), 630–633 (1998)
3. Biere, A., Cimatti, A., Clarke, E., Zhu, Y.: Symbolic Model Checking without BDDs. In: Tools and Algorithms for the Construction and Analysis of Systems, vol. 1579 of LNCS. 1999, pp. 193–207 (1999)
4. Cavada, R., et al.: The nuXmv symbolic model checker. In: Proceedings of CAV (2014), vol. 8559 of LNCS, pp. 334–342 (2014)
5. Department of Education and Training, State of Victoria. High impact teaching strategies: excellence in teaching and learning. ISBN: 978-0-7594-0820-3 (2020)
6. Ferguson, P.: Student perceptions of quality feedback in teacher education. Assess. Eval. High. Educ. **36**(1), 51–62 (2011)
7. Ghilay, Y., Ghilay, R., et al.: Tbal: technology-based active learning in higher education. J. Educ. Learn. **4** (9 2015)
8. Howell, R.A.: Engaging students in education for sustainable development: the benefits of active learning, reflective practices and flipped classroom pedagogies. J. Cleaner Product.**325** (11 2021)
9. Jackson, D.: Software Abstractions: logic, language, and analysis. MIT press (2012)
10. Laal, M., Ghodsi, S.M.: Benefits of collaborative learning. Procedia. Soc. Behav. Sci. **31**, 486–490 (2012)
11. Li, Y.W.: Transforming conventional teaching classroom to learner-centred teaching classroom using multimedia-mediated learning module. Int. J. Inform. Educ. Technol. **6**, 105–112 (2016)
12. Liu, M., et al.: Creating a multimedia enhanced problem-based learning environment for middle school science: Voices from the developers. Interdiscip. J. Problem-Based Learn. **8** (3 2014)
13. Macedo, N., et al.: Experiences on teaching alloy with an automated assessment platform. Sci. Comput. Program. **211**, 102690 (2021)
14. María-Del-Mar Gallardo, L. P.: Modelling and specifying software systems with alloy * (tutorial)
15. Moreira, R.M., Paiva, A.C.: A novel approach using alloy in domain-specific language engineering. In: 2015 3rd International Conference on Model-Driven Engineering and Software Development (MODELSWARD) (2015), IEEE, pp. 157–164 (2015)
16. Pnueli, A.: The temporal logic of programs. In: 18th Annual Symposium on Foundations of Computer Science (sfcs 1977) (1977). IEEE, pp. 46–57 (1977)
17. Renkl, A.: The Worked-Out Examples principle in Multimedia Learning. **01**, 229–245 (2005)

18. Seidel, T., Rimmele, R., Prenzel, M.: Clarity and coherence of lesson goals as a scaffold for student learning. Learning and Instruction - LEARN INSTR 15 (12 2005), 539–556 (2005)
19. Smith, A.R., Cavanaugh, C., Moore, W.A.: Instructional multimedia: an investigation of student and instructor attitudes and student study behavior. BMC Med. Educ. 11 (2011)
20. van Merrienboer, J.: Training complex cognitive skills: a four-component instructional design model for technical training. Educational Technology Publications (1997)

Checking Contracts in Event-B

Reporting the Introduction and the Use of Automated Tools for Verifying Software-Based Systems in Higher Education

Dominique Méry[✉] [iD]

LORIA & Université de Lorraine, Vandœuvre-lès-Nancy, France
dominique.mery@loria.fr
http://members.loria.fr/Mery

Abstract. Verification of program properties such as partial correctness (PC) or absence of errors at runtime (RTE) applies induction principles using algorithmic techniques for checking statements in a logical framework such as classical logic or temporal logic. Alan Turing was undoubtedly the first to annotate programs, namely Turing machines, and to apply an induction principle to transition systems. Our work is placed in this perspective of verifying safety properties of programs which could be executed sequentially or in a distributed manner, with the aim of presenting them as simply as possible to student classes in the context of a posteriori verification. We report on an in vivo experiment using the Event-B language and associated tools as an assembly and disassembly platform for correcting programs in a programming language. We have adopted a contract-based approach to programming, which we are implementing with Event-B . A few examples are given to illustrate this pedagogical approach as well as comments and observations. This step is part of a process of learning both the underlying techniques and other tools such as Frama-C based on the same ideas.

Keywords: Verification · Safety · Program Properties · Event-B

Supported by the ANR DISCONT Project (ANR-17-CE25-0005) and by the ANR EBRP Plus Project (ANR-19-CE25-0010). The file is generated on August 2, 2024.

E. Sekerinski and L. Ribeiro (Eds.): FMTea 2024, LNCS 14939, pp. 91–105, 2024.
https://doi.org/10.1007/978-3-031-71379-8_6

1   Introduction

Programming by contract, as outlined in Meyer's work [22], is based on a *contract* between the software developer and software user. In Meyer's terms, the supplier and the consumer are linked by a contract, which expresses a link between a pair (precondition, postcondition) and a possibly annotated algorithm. The objective is to utilise the Event-B modelling language [2] as a framework for expressing verification conditions for contracts and to compare the resulting Rodin-based tool to other existing automated verification tools, such as Frama-C [6]..This exercise also introduces the Event-B language and the use of the Rodin [3] and Atelier-B [1,5,9] environments.

Our current work is related to our lectures on *modelling, designing, verifying and validating software-based systems* taught in the MsC *Computer Science* at Faculty of Science of the University of Lorraine and in the Computer Engineering Master of the School *Telecom Nancy* of the University of Lorraine. The epistemological concepts were given using the classical blackboard and chalk and progressively we have moved to integrate automated verification techniques and tools such as Dafny [14], Why3 [23] and Frama-C [6]. Our list is not exhaustive and our idea is to introduce progressively the concepts of verification using the Floyd-Hoare principle and to show how students can develop a tool for their pet programming language.

Our main reference is the work of Patrick and Radhia Cousot [12] who analyse the (16) different induction principles for proving program invariance properties. Figure 1 sketches the main steps of our method:

- FORMALISATION Expression of the contract as assertions defined in an Event-B context.
- TRANSLATION Translation of annotations as elements of the invariant and of the basic computation steps between two successive labels as events.

The SEES clause is implemented in the Event-B modelling language by the Rodin platform.

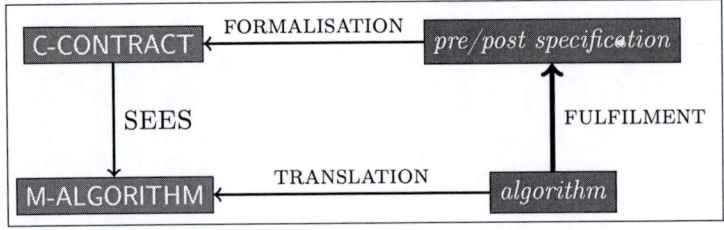

Fig. 1. The verification pattern

In [7], authors are describing a translation in a different way. They use the Dafny tool for checking the Event-B context and machines. In our case, it is a matter of replacing Dafny and making full use of Rodin and its associated provers. This is an important exercise for students learning the Event-B notation, who will then be able to put the question of refinement into practice. It is also a way to illustrate the use of proof checkers when dealing with proofs of program properties. Finally, we found the way to write the proof checking using Event-B and Rodin without toil.

Before we present the technical and formal elements, we must set out the framework for this experiment. Its aim is to introduce formal methods into the academic curriculum and train students in their use. The rest of the paper is organised as follow.

Section 2 gives the academic context. Section 3 introduces Event-B notations required for expressing contracts; we give a short description of the small programming language used for illustrating concepts. Section 4 introduces basic notations and concepts for programming by contracts, the translation of a contract into an Event-B context and a Event-B machine. Section 5 is giving several examples of contract for classical algorithms with comments on the proof process. Finally we conclude and give some perspectives. The full version of the paper with Rodin archives is available at the link[1].

2   Academic Context

Our experimentation is long and began during the academic year 1983/1984 at the University of Metz. As assistant professor, I taught tutorials on the Floyd-Hoare method, including the treatment of sequential programming language concepts like pointers, functions, procedures, etc. As professor, Patrick Cousot was my colleague and taught lectures on the same topics. Then our paths diverged. From 1983 to 1993, I taught the same course as Patrick Cousot, who joined the Ecole Polytechnique in 1984. I led a project based on the publications of Clarke, Emerson and Sistla [8]. In this project, students implemented the method from this publication. One of the projects carried out in Pascal on a Micral PC running DOS was a resounding success. It was possible to verify temporal logic properties on finite transition systems. However, I remained fascinated by the general case and finite models seemed limited.

At the time, we had created a verification tool for SDL programs in the CONCERTO environment using the young ISABELLE prover [21]. This experience convinced me of the usefulness of the tools and the feasibility of such an approach. My appointment as a professor at the University of Nancy 1 in 1993 placed me once again in a new context: that of a computer science engineering school. I was given the course called Models and Algorithms (MALG), which covered computability, complexity, fixed-point theory (Kleene), propositional calculus, first-order calculus, resolution, verification of algorithms and functional programming. It had an hourly volume of 48 h 00 for lectures and 48 h 00 for tutorials. From 1988 to 1996, formal modelling languages were the focus of my research activities and teaching on formal methods. The CROCOS

---
[1] https://mery54.github.io/fmt/.

experiment [21] proposed a model of systems as a set of conditions/actions and a linear temporal logic.

My teaching history has led me to make some important observations that will inform the development of my teaching methods for formal methods: (1) It is clear that the MALG teaching programme was not sufficiently based on tools and projects; (2) The concepts linked to the semantics and logic of programs must be put in direct relation with the uses of programming features; (3) A student of computer science must manage models of different types; and (4) modelling remains an activity based on practice and the use of tools.

The MALG course is the first formal introduction to computer engineering training. It is taken in the second year of the course. Students have already received basic training in computer science in the first year, including mastery of a programming language, computer architecture and databases. The second year builds on this and leads to a third year during which a formal modelling course using the Event-B language is given. MALG is therefore a preparation for this 24-hour formal modelling course. Lastly, the school's students are recruited after a highly selective competitive examination, following two years of preparation in preparatory classes for entry to engineering schools. They are awarded a Master's degree in computer science at the end of their three years at the school. The number of students enrolled on the MALG course is between 50 and 70; the number of third-year students on the formal modelling course is between 25 and 30.

The students have a good level of science but have chosen computer science to develop their careers. By learning the basics of computer science, they are made aware of the issues surrounding bugs in programs. However, a large number of comments indicate that they are often unconcerned about run-time errors, the storage of information, and issues related to addresses in programs. It is clear that modern programming languages have been able to protect users from certain pitfalls associated with function or procedure calls and parameter passing mechanisms. Amongst the students, one group is aiming to go further into the design of embedded systems, using low-level computing techniques; my colleagues leave me free to choose the languages and tools, but it is important to develop skills in model checking. Among the languages, there is the C language used by Frama-c in an extended but ACSL-compatible form, as well as LUSTRE, which was chosen to introduce the synchronous hypothesis, with tools from the VERIMAG platform [17], as well as using the KIND2 tool [24].

In the first lesson we use examples of C programs to justify what we are going to introduce. The C program of listing 1.1 of Fig. 2, for example, returns an unexpected value for almost all the students. Only one or two students explain and predict the result. We then look at tests, which are still not very well mastered; we then give an example of a non-testable program in the listing 1.2, and show how Frama-c reacts by suggesting a condition. We give the sequence of programs that we have developed from these two examples in the appendix to the complete version (see https://mery54.github.io/fmt/) of this document.

Listing 1.1: Function average

```
#include <stdio.h>
#include <limits.h>
int average(int a,int b)
{
    return((a+b)/2);
}

int main()
{
    int x,y;
    x=INT_MAX;y=INT_MAX;
    printf("Average__for_%d
____and_%d_is_%d\n",x,y,
            average(x,y));
    return 0;
}
```

Listing 1.2: Non testable Function

```
#include <stdio.h>
#include <stdlib.h>
#include <time.h>

int main() {
    int x, y;
    // Seed the random number
    // generator
    // with the current time
    srand(time(NULL));
    // Generate a random number
    // between 1 and 100
    x = rand() % 100 + 1;
    // Perform some calculations
    y = x / (100 - x);
    printf("Result:_%d\n", y);
    return 0;
}
```

Fig. 2. Two C programs for lectures

In our paper, we will focus on the technical concepts that we have used to model programs, but also on program properties and on the available software tools that implement analysis, simulation and proof techniques. We will also comment on the tools used, which are not limited to Rodin but are used jointly and in a complementary way.

To conclude this presentation of the academic context, we set out our pedagogical objectives in relation to the targeted training course, which concerns the design of embedded or non-embedded software systems with modelling, verification and validation. The challenge is to teach the mastery of software modelling and its properties and the use of verification techniques, in particular mathematical proof.

3   Modelling and Programming Languages

3.1   Summary on Event-B

Event-B is a correct-by-construction, stated-based formal modelling language for system design [2]. First-order logic (FOL) and set theory underpin the Event-B modelling language. The design process consists of a series of refinements of an abstract model (specification) leading to a final concrete model. Refinement progressively contributes to add design decisions to the system. We are not considering the refinement relation in this paper. Three components define Event-B models: *Contexts*, *Machines*, and *Theories*. However, we will not use *Theories* [19] and will not describe this concept.

A *Context* (Fig. 3) is the static part of a model. It is used to set up definitions, axioms, and theorems needed to describe required concepts. *Carrier sets s* defining algebraically new types (possibly constrained in axioms or other extending contexts), *constants c*, *axioms* $AX(s,c)$ and *theorems* $TH(s,c)$ are introduced.

A *machine* (Fig. 3) describes the dynamic part of a model as a transition system. A set of possibly parameterised and/or guarded events (transitions) modifying a set of state variables (state) represents the core concepts of a machine.

MACHINE
  a
REFINES
  am
SEES
  co
VARIABLES
  x
INVARIANTS
  $I(s,c,x)$
THEOREMS
  $S(s,c,x)$
VARIANT
  $V(s,c,x)$
EVENTS
INIT
  BEGIN
    $x : | (IP(s,c,x'))$
  END $\ldots$
e
  ANY α
  WHERE
    $G(s,c,\alpha,x)$
  THEN
    $x : | (P(s,c,\alpha,x,x'))$
  END END

CONTEXTS
  co
EXTENDS
  aco
SETS
  s
CONSTANTS
  c
AXIOMS

$$AX(s,c) = \left[\begin{array}{l} \ldots \\ ax : ax(s,c) \\ \ldots \end{array} \right.$$

THEOREMS

$$TH(s,c) = \left[\begin{array}{l} \ldots \\ th : th(s,c) \\ \ldots \end{array} \right.$$

END

(1) $Theorems(AX)$
  $AX(s,c) \vdash th(s,c)$
(2) $Theorems(TH)$
  $AX(s,c), I(s,c,x) \vdash S(s,c,x)$
(3) $Initialisation(INIT)$
  $AX(s,c), IP(s,c,x') \vdash I(s,c,x')$
(4) $Invariant(INV)$
  $AX(s,c), I(x),$
$G(s,c,\alpha,x), P(s,c,\alpha,x,x')$
  $\vdash I(s,c,x')$
(5) $Feasibility(FIS)$
  $AX(s,c), I(s,c,x), G(s,c,\alpha,x)$
  $\vdash \exists x' \cdot P(s,c,\alpha,x,x')$

Fig. 3. Event-B structures: Context & Machine and Proof Obligations

Variables x, *invariants* $I(s,c,x)$, *theorems* $S(s,c,x)$, *variants* $V(x)$, and *events* Event e (possibly guarded by G and/or parameterised by α) are defined in a machine. *Invariants* and *theorems* formalise system safety properties while *variants* define convergence properties (reachability).

Before-After Predicates (BAP) express state variables changes using prime notation x' to record the new value of a variable x after a change. The *"becomes such that"* $:|$ substitution is used to define the next (transition or event) value of a state variable. We write $x :| P(s,c,\alpha,x,x')$ to express that the next value of x (denoted by x') satisfies the predicate $P(s,c,\alpha,x,x')$ defined on before and after values of variable x. When a parameter α is involved in a variable the BAP is expressed as $x :| P(s, c, \alpha, x, x')$.

To establish the correctness of an Event-B machine, POs (automatically generated from the calculus of substitutions) need to be proved.

The main proof obligations (POs), relevant for this paper, are listed in the table of Fig. 3. They require to demonstrate the context and machine theorems (1,2), initialisation (3), invariant preservation (4) and event feasibility (5).

Rodin[2] is an open source, Eclipse-based Integrated Development Environment for modelling in Event-B . It offers resources for model editing, automatic PO generation, project management, refinement, proof, model checking, model animation, and code generation. Event-B theories extension is available in the form of a plug-in, developed for the Rodin platform. Many provers like predicate provers, SMT solvers, are plugins for Rodin.

Comments and observations 1. *The main difficulty in modelling with Event-B is that the modelling language is very abstract and the notation $x :$ $|R(s, c, x, x')$ is very general. Set theory allows a certain amount of freedom in modelling, in particular to express relationships between values that are not always computable. For example, an Event-B variable corresponds to a state of an algorithm or an observed program, but this variable is by nature perdurant or flexible; it has a current value, an initial value and sometimes a final value. The term "variable" is in fact very overloaded and a distinction must be made between these perdurable variables and the variables of logical formalims. We separate the initial values from the variables by deciding to use an index 0 or f, where x0 is the initial value of x and xf is the final value of x. The ACSL language can be used to designate variable values at given points, and it is important to unpack or dissect these concepts before using them effectively. The introduction of Event-B was in response to a comment made by students who were first introduced to TLA+ using the ToolBox (or VSCode) platform and who had to write lists of definitions that were tested using the platform's model checker. The PlusCal language has made the use of TLA concepts more transparent, but our objective is still to learn a language in two stages: firstly, to check or describe algorithms, and then to use it as a modelling language, but with model correctness through refinement. The aim is to train students to manipulate formal concepts using IT tools.*

3.2   Programming Constructs

Programming constructs are classical constructs as assignment ($v := f_{ell,\ell'}(v)$), skip statement skip, conditional statement (if cond(v) S_1 else S_2 fi) and iterative statement (while cond(v) do S od). We use these constructs for expressing programs or algorithms which are annotated possibly by labels.

---

[2] http://www.event-b.org/index.html.

ℓ_0 :
$k := 0$;
ℓ_1 :
$co := 0$;
ℓ_2 :
while $(k < n)$ do
   ℓ_3 :
     if $(k \,\%2 \,== \,0)$
   ℓ_4 :
       $co := co + k + 1$;
     fi;
   ℓ_5 :
     $k := k + 1$;
od;
   ℓ_6 :
$ro := co$;
   ℓ_5 :

There different ways to annotate algorithms. One can assign a label $\ell \in L$ to each statement:

- $\mathsf{v} := \mathsf{f}_{ell,\ell'}(\mathsf{v})$ is labelled as follow
  ℓ :
     $\mathsf{v} := \mathsf{f}_{\ell,\ell'}(\mathsf{v})$;
- if $\mathsf{cond}(\mathsf{v})$ S_1 else S_2 fi is labelled as follow
  ℓ :
     if $\mathsf{cond}(\mathsf{v})$ S_1 else S_2 fi
- while $\mathsf{cond}(\mathsf{v})$ do S od is labelled as follow
  ℓ :
     while $\mathsf{cond}(\mathsf{v})$ do S od

The annotation process uses one label at most one time. For instance, the following annotation of a small algorithm is a small example in the left box.

Each pair of successive labels ℓ, ℓ' is interpreted by a condition denoted $\mathsf{cond}_{\ell,\ell'}(v)$ and an assignment $\mathsf{v} := \mathsf{f}_{\ell,\ell'}(v)$. A flowchart can be derived following the next diagram:

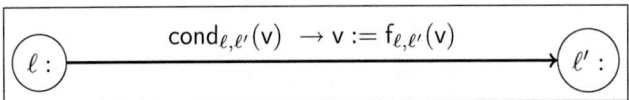

In our paper, we assume that the programming language is $\mathcal{PL}$ and one can derive a flowchart from the annotated algorithmic notation.

Comments and observations 2. *Initially we used examples borrowed from Manna [20], which uses flowcharts and a structured language to describe algorithms. This book is very comprehensive and gives an overview of important topics in theoretical computer science, and is still useful for students. The course consists of translating some flowcharts into TLA notation by hand, then doing the same with algorithms in structured form, and finally experimenting with Plus-Cal. The course therefore starts by gradually learning how to translate a model and then to check the model obtained for correctness properties such as partial correctness or absence of errors at runtime. During this phase, students learn the relationship between non-primed and primed variables, as well as set notation. Students first discover TLA/TLA+ through the ToolBox tool [16] , which allows them to write state transformations in a modelling language that includes set theory. The students quickly understood the limitations of the model checking tool and the advantages of parametrisation by a constant. It's a question of convincing with effective tools, and it's clear that the Rodin tool hides the induction stages with a list of proof obligations, which we'll come back to. Our idea is also to train students in the design of a method for proving program properties for their programming language.*

It could be argued that the TLA+ ToolBox platform [16] provides a powerful and effective proof tool based on Isabelle/HOL and SMT solvers, but the problem is that its use requires a greater effort on the part of the student, who has to manage the generation of verification conditions himself. In fact, the Event-B language describes a model which must be inductive and which describes the observations of changes of state of the variables in the model.

4   Programming by Contract

Programming by contract [22] is based on a contract between the software developer and software user - in Meyer's terms the supplier and the consumer. Every process starts with a precondition that must be satisfied by the consumer and it ends with postconditions which the supplier guarantees to be true (if and only if the precondition was met). The contract is defined by two assertions a precondition and a postcondition; the algorithm is annotated. The postcondition establishes a relation between the initial values of variables and the final values of variables.

We use two languages a programming or algorithmic language $\mathcal{PL}$ for expressing algorithms and an assertion language denoted $\mathcal{AL}$ for expressing annotations. A contract is a pair $(pre(v_0), post(v_0, v_f)$ where $pre(v_0)$ states the specification of input values denoted v_0 and v_0 is the initial value of the variable v. $post(v_0, v_f)$ is the relation between the initial values v_0 of v. and the final values of v.

We adopt a convention to make our explanation as clear as possible and we will denote *non-logical* (or computer or flexible [18]) variables by strings using the font as v, Tax, Result, ... and logical variables by strings using the font as v, Tax, $Result$, ... The convention is adapted from Patrick Cousot's comments [10] on making a distinction between a value of a computer variable and the computer variable itself.

A program or an algorithm P over variables v *fullfills* a contract $(pre(v_0), post(v_0, v_f)$, when:

- P transforms a non-logical variable v from an initial value v_0 to a final value
v_f: $v_0 \xrightarrow{\text{P}} v_f$
- v_0 satisfies pre: $pre(v_0)$ and v_f satisfies a relation post : $post(v_0, v_f)$
- $pre(v_0) \wedge v_0 \xrightarrow{\text{P}} v_f \Rightarrow post(v_0, v_f)$

We will denote a *contract* for P as follows.

```
contract P
variables v
requires pre(v_0)
ensures post(v_0, v_f)
```

The contract has a name which is the name of the program under construction. That program may be implicit or explicit. It may be a program which is not yet existing and we may follow the *refinement*-based approach or a direct construction.

As pointed out by C. Jones in his speech accepting the FM fellow, a postcondition is a relation between the current value of variables and their initial values. P.

and R. Cousot [11, 12] give detail on induction principles of the proposed methods as Hoare, Manna ... and partition invariance proof methods into assertional ones and relational ones; they explain how they are related using a cube representation and Galois connections for expressing these relationships. We consider the following general interpretation of $P(x)$ by expressing it as $x \in \tilde{P}$ from a correspondance between a predicate and the set of values validating this predicate. For ease of syntax we leave out the $\tilde{}$ symbol.

Comments and observations 3. *We highlight the elements that characterise a contract and emphasise the importance of the initial and final values of flexible variables. Students will be introduced to the question of **what** to compute and what not to compute. The post-condition is a relationship between initial and final values. A distinction is made between logical and non-logical variables. Finally, it is important to justify the link between this contract and the algorithm in question (the **how**). This stage justifies the principles of induction and highlights the theory of fixed-points on complete lattices. We emphasise the notion of computability associated with the algorithm. A set of verification conditions is derived from this induction and used to verify the annotations. At this point we talk about the connection with the strongest invariant and we obtain the Floyd-Hoare verification conditions.*

The translation in Rodin is simple and we have to define the domain of variables namely D. We have chosen a general form. The context is used for expressing theorems required for deriving the postcondition. The context SQUARE-C0 corresponds to the contract for computing the square of a positive integer. In this case, we have to define a sequence which is supporting the computation of the square of a natural number.

```
CONTEXT C0
SETS
    D
CONSTANTS
    v0, vf, post, pre
AXIOMS
    def1 : pre ⊆ D
    def2 : post ⊆ D × D
    pre(v0) : v0 ∈ pre
    post(v0, vf) : v0 ↦ vf ∈ post
END
```

Example 41. *Contract in Event-B for square computation*

```
CONTEXT SQUARE − C0
CONSTANTS
    n0, r0, nf, rf
AXIOMS
    pre(n0, r0) : n0 ∈ ℕ ∧ r0 ∈ ℤ
    post(n0, r0, nf, rf) :  nf = n0
                            rf = n0 * n0
END
```

The contract **SQUARE** is expressing the relation of computation of the square of n.

contract SQUARE
variables n,r
requires $n0 \in \mathbb{N} \land r0 \in \mathbb{Z}$
ensures $nf = n0$ $rf = n0 * n0$

A contract can be extended by the definition of an algorithmic section which is describing the computation process itself. The annotation of the algorithmic section is not required but it can help the proof process and it will be generally checked using *verification conditions* following the *Floyd-Hoare method* [13,15]. The contract is stated and a code is added.

Verification conditions are listed as follows:

```
contract P
variables v
requires pre(v0)
ensures post(v0, vf)
  ⌈ begin
  │ 0 : P0(v0, v)
  │ S0
  │ ...
  │ i : Pi(v0, v)
  │ ...
  │ Sf−1
  │ f : Pf(v0, v)
  ⌊ end
```

- (initialisation)
  $$pre(v_0) \wedge v = v_0 \Rightarrow P_0(v_0, v)$$
- (finalisation)
  $$pre(v_0) \wedge P_f(v_0, v) \Rightarrow post(v_0, v)$$
- (induction)
  For each labels pair ℓ, ℓ'
  such that $\ell \longrightarrow \ell'$, one checks that,
  for any value $v, v' \in D$
  $$\left(\begin{pmatrix} pre(v_0) \wedge P_\ell(v_0, v)) \\ \wedge cond_{\ell,\ell'}(v) \wedge v' = f_{\ell,\ell'}(v) \end{pmatrix} \Rightarrow P_{\ell'}(v_0, v') \right),$$

Three kinds of verification conditions should be checked and we justify the method in the full version..

The method checks that the annotation denoting verification is correct. An Event-B machine (see Fig. 4) is built from the extended contract.

```
MACHINE M
SEES    C0
VARIABLES
   v, pc
INVARIANTS
   typing : v ∈ D
   control : pc ∈ L
   ...
   atℓ : pc = ℓ ⇒ Pℓ(v0, v)
   ...
th1 : pre(v0) ∧ v = v0 ⇒ P0(v0, v)
th2 : pre(v0) ∧ Pf(v0, v)
              ⇒ post(v0, v)
...
END
...
END
```

```
MACHINE M
EVENTS
INITIALISATION
BEGIN
  (pc, v) : | ( pc' = l0 ∧ v' = v0
              ∧pre(v0)         )
END
...
e(ℓ, ℓ')
   WHEN
       pc = ℓ
       condℓ,ℓ'(v)
   THEN
       pc := ℓ'
       v := fℓ,ℓ'(v)
   END
...
END
```

Fig. 4. Event-B machine for checking contract

The machine M (Fig. 4) has variables as v for modelling v and we add a control variable pc whose values are in L. For each label ℓ, one adds an implication defining the current state, when, the control is at ℓ. The initialisation of the variables is defined by the precondition and the initial possible values $v0$. Events are defined for each pair of labels (ℓ, ℓ') and is modelling the flowchart derived from the algorithm. In Sect. 5, we give examples which illustrate the methodology.

Comments and observations 4. *We translated the verification conditions derived from our strongest invariant given by the fixed-point definition. Then we derived a set of annotations for verifying the algorithm against the contract, but this is just a translation of what we've been doing for many years and it came across as very natural and immediate during a tutorial. The idea was to try this translation and this communication translates this episode. But from a contract point of view, we can use the ACSL language, which can be used to express all the above concepts and which is equipped with a Frama-c verification tool with the wp plugin. We could stop there, but the story continues. In fact, the wp plugin uses the wp calculation to generate the verification conditions. This leads us to introduce the students to the wp calculation in its two forms, depending on whether we are considering total or partial correctness. It turns out that the wp plugin analyses the contract according to the Hoare logic system and we must therefore show the equivalence of the two approaches Floyd-Hoare and wp. In the short version, we omit the proof of correctness of our method but the students' understanding depends on the use of the wp calculus by hand. The work of Patrick and Radhia Cousot [12] is clear and we refer the reader to Patrick Cousot's book [10] which explains the relationship between these two ways of checking an algorithm against its contract. We have simply taken up the concepts and disseminated them over the past few years.*

5   The Methodology in Action

We have described concepts required for using Event-B as support for expressing verification conditions that have been given. We justify verification conditions in the full version . This example does not illustrate the prover's performance, but rather the simplicity of the translation. We developed this translation based on a tutorial session during which I tested it without proving its correctness. We wanted to check the manual verification process, which works by applying simple rewriting and sequence simplification rules. The method allows you to teach the Event-B language and to state the need of refinement.

Obtaining the invariant simply involves copying annotations as a conjunction of local annotations: invariants inv1 and inv2 are type invariants, inv3, inv4 and inv5 come from the contract SIMPLE. x_0 is the input value of x and x_f is the final value of x at ℓ_2.

contract SIMPLE
variables x
requires $x_0 \in \mathbb{N}$
ensures $x_f = 0$
begin
$\ell_0 : \{0 \leq x \leq x_0 \land x_0 \in \mathbb{N}\}$
while $0 < \mathrm{x}$ **do**
$\quad \ell_1 : \{0 < x \land x \leq x_0 \land x_0 \in \mathbb{N}\}$
$\quad \mathrm{x} := \mathrm{x} - 1;$
od
$\ell_2 : \{x = 0\}$end

INVARIANTS
$inv1 : x \in \mathbb{N}$
$inv2 : l \in L$
$inv3 : l = l0 \Rightarrow$
$\quad 0 \leq x \land x \leq x0 \land x0 \in \mathbb{N}$
$inv4 : l = l1 \Rightarrow$
$\quad 0 < x \land x \leq x0 \land x0 \in \mathbb{N}$
$inv5 : l = l2 \Rightarrow x = 0$
$requires : x0 \in \mathbb{N} \land x = x0$
$\Rightarrow x = x0 \land x0 \in \mathbb{N}$
$ensures : x = 0 \land x = x0$
$\Rightarrow x = 0$

The writing process is straightforward for students. They write an invariant and then the events corresponding to the observation of the calculation described by the algorithm. Students concentrate mainly on the formal writing of the annotations and only discover the result of the proof when the file is saved.

Event $el0l2$
WHEN
$\quad grd1 : l = l0$
$\quad grd2 : \neg(0 < x)$
THEN
$\quad act1 : l := l2$

Event $Init$
THEN
$\quad act1 : x := x0$
$\quad act2 : l := l0$

Event $el0l1$
WHEN
$\quad grd1 : l = l0$
$\quad grd2 : 0 < x$
THEN
$\quad act1 : l := l1$

Event $el1l0$
WHEN
$\quad grd1 : l = l1$
THEN
$\quad act1 : l := l0$
$\quad act2 : x := x - 1$

This example is very simple and consists of one iteration which stops when the value of x is zero. It does not pose problem with Rodin and the proofs are derived at the same time as the at the same time as writing the elements of the invariant and the events. Proofs obligations are discharged by the proof tools while editing the Event-B machine in the Rodin platform.

6   Final Comments and Conclusion

The evolution of teaching in our courses on software engineering and distributed algorithms is marked by the use of a number of verification tools with master level students. In fourth year, students learn the basic concepts and techniques they will need to know, including how to use logic to model program properties, the semantics of programming languages and induction principles. They are trained in fundamental tools that they will almost certainly need to use, such as model checking, runtime verification or test management. Additionally, we sought to collaborate with students on authentic programming challenges, which led to the development of a language centered around contracts.

Upon their arrival in the fourth year, we realized that there was a lack of connection between the problem posed by a particular execution, such as calculating the average of two numbers in C, and the proposal to calculate this average for the two numbers equal to the maximum that can be coded. The value returned (-1) is still largely misunderstood. Using Frama-C demonstrated that the RTE plugin facilitated the management of potential errors. The most common issues are managing tools and, in particular, distributing them across different types of operating systems. One solution is to create a virtual machine with the necessary software installed, but this can cause problems on machines that are not powerful enough or too new.

The second stage of our project to integrate formal methods into a university curriculum is to teach the Event-B modelling language and to use incremental development based on refinement. We will handle the notion of contract in 4th year so that we can continue to master the Event-B language and, in particular, to introduce the refinement of formal models. The MALG course are reviewed with the formal expression of refinement and its use. In particular, the incremental development of sequential and distributed algorithms is covered with Event-B and Rodin. The leader election algorithm [4]. was the starting point for this work. It made it possible to explain this algorithm simply to students. Our students cohorts include a significant proportion of students who have learned mathematical proof while preparing for university entrance examinations. These students are well-equipped to play with the tools and interact effectively. Finally, we would like to emphasise the Knaster-Tarski [10] theorem, which also allows us to play with inductions and inductive definitions. We have employed the work of P. and R. Cousot [12] to define an induction principle for proving safety properties in the case of sequential programs, which can be generalised to concurrent or distributed programs. This translation demonstrates how verification tools operate and illustrates the link between the semantics of programming languages and the verification process.

References

1. Abrial, J.-R.: The B-Book - Assigning Programs to Meanings. Cambridge University Press (1996)
2. Abrial, J.-R.: Modeling in Event-B: System and Software Engineering. Cambridge University Press (2010)
3. Abrial, J.R., Butler, M., Hallerstede, S., Hoang, T.S., Mehta, F., Voisin, L.: Rodin: an open toolset for modelling and reasoning in Event-B. STTT, 12(6), 447–466 (2010). https://doi.org/10.1007/s10009-010-0145-y
4. Abrial, J.-R., Cansell, D., Méry, D.: A mechanically proved and incremental development of IEEE 1394 tree identify protocol. Formal Aspects Comput. 14(3), 215–227 (2003)
5. Barradas, H.-.: Event-B: syntax and proof oglogations in Atelier-B. Technical report, ClearSy (2020)
6. Baudin, P., et al.: The dogged pursuit of bug-free C programs: the Frama-c software analysis platform. Commun. ACM 64(8), 56–68 (2021)

7. Catano, N., Leino, K.R.M., Rivera, V.: The eventb2dafny Rodin plug-in. In: Garbervetsky, D., Kim, S., (eds.) Proceedings of the Second International Workshop on Developing Tools as Plug-Ins, TOPI 2012, Zurich, Switzerland, June 3, 2012, pp. 49–54. IEEE Computer Society (2012)

8. Clarke, E.M., Emerson, E.A., Sistla, A.P.: Automatic verification of finite state concurrent systems using temporal logic specifications: a practical approach. In: Wright, J.R., Landweber, L., Demers, A.J., Teitelbaum, T. (eds.) Conference Record of the Tenth Annual ACM Symposium on Principles of Programming Languages, Austin, Texas, USA, January 1983, pp. 117–126. ACM Press (1983)

9. ClearSy. B Language reference manual ver.1.8.10 (2022)

10. Cousot, P.: Principles of Abstract Interpretation. The MIT Press (2021)

11. Cousot, P.: Calculational design of [in]correctness transformational program logics by abstract interpretation. Proc. ACM Program. Lang. **8**(POPL), 175–208 (2024)

12. Cousot, P., Cousot, R.: Induction principles for proving invariance properties of programs. In: Néel, D. (ed.) Tools & Notions for Program Construction: an Advanced Course, pp. 75–119. Cambridge University Press, Cambridge, UK (1982)

13. Floyd, R.W.: Assigning meanings to programs. In: Colburn, T.R., Fetzer, J.H., Rankin, T.L. (eds.) Program Verification: Fundamental Issues in Computer Science, pp. 65–81. Springer, Netherlands, Dordrecht (1993). https://doi.org/10.1007/978-94-011-1793-7_4

14. Ford, R.L., Leino, K.R.M.: Dafny reference manual (2017)

15. Hoare, C.A.R.: An axiomatic basis for computer programming. Commun. ACM **12**(10), 576–580 (1969)

16. Kuppe, M.A., Lamport, L., Ricketts, D.: The TLA+ toolbox. In: Monahan, R., Prevosto, V., Proença, J. (eds.) Proceedings Fifth Workshop on Formal Integrated Development Environment, F-IDE@FM 2019, Porto, Portugal, 7th October 2019. EPTCS, vol. 310, pp. 50–62 (2019)

17. Verimag Laboratory: the synchrone reactive toolbox (2022). https://www-verimag.imag.fr/DIST-TOOLS/SYNCHRONE/reactive-toolbox/

18. Lamport, L.: The temporal logic of actions. ACM Trans. Program. Lang. Syst. **16**(3), 872–923 (1994)

19. Maamria, I., Fathabadi, A.S.: Theory Plug-in User Manual. University of Southampton, 30 April 2014

20. Manna, Z.: Mathematical Theory of Computation. Dover Publications Inc, US (2003)

21. Méry, D., Mokkedem, A.: Crocos: an integrated environment for interactive verification of SDL specifications. In: von Bochmann, G., Probst, D.K. (eds.) CAV 1992. LNCS, vol. 663, pp. 343–356. Springer, Heidelberg (1993). https://doi.org/10.1007/3-540-56496-9_27

22. Meyer, B.: Design by contract. In: Mandrioli, D., Meyer, B. (eds.) Advances in Object-Oriented Software Engineering, pp. 1–50 (1991)

23. Why3 Team. Why3. https://why3.lri.fr

24. Iowa University. Kind2 multi-engine SMT-based automatic model checker for synchronous reactive systems (2024). https://kind2-mc.github.io/kind2/

Teaching with Logika: Conceiving and Constructing Correct Software

Stefan Hallerstede[1][(✉)], John Hatcliff[2], and Robby[2]

[1] Aarhus University, Aarhus, Denmark
sha@ece.au.dk
[2] Kansas State University, Manhattan, USA
{hatcliff,robby}@ksu.edu

Abstract. Slang is a subset of Scala designed for coding high assurance software. It is supported by a highly automated verification tool called Logika that incorporates multiple forms of formal methods, presented in terms of programming-oriented notations and activities. In this paper, we describe our teaching approach for how to conceive and construct correct software using Slang and Logika. A key feature of our approach includes presenting specification, coding, testing, and verification as an integrated engineering methodology. We further motivate students by illustrating Slang/Logika's use in broader contexts of model-based development and assurance of critical embedded systems. We describe our experience in using Logika for teaching at both the undergraduate and graduate level. A variety of pedagogical materials are available including the open source Slang/Logika implementation, lecture slides, and course notes.

1 Introduction

Teaching formal methods is a challenging task. Sometimes students struggle to understand the foundations, and they are not convinced that a formal method will be useful in practice. This often leads to a primary pedagogical focus on communicating the semantics of a technique and drawing (unfavorable) contrasts with informal methods. As a community, we have come to understand that students absorb and apply formal methods concepts best when they are tool supported, yet this sometime creates a problem that students do not understand what the tool does, and they may not be able to transfer useful reasoning concepts into settings where they do not have formal methods tools or where they need to work with informal methods. Furthermore, fellow faculty members are often not sympathetic to the use of formal methods, so often times learning and applying formal methods is confined to a single course or perspective, and students fail to see how rigorous reasoning can be applied throughout system development and across different domains.

At Aarhus University (AU) and Kansas State University (KSU), we are pursuing a curriculum approach that emphasizes *integrating formal and informal methods* in the context of an *engineering discipline* that *spans multiple levels of student learning* (e.g., both undergraduate and graduate) and is demonstrated across *many phases of system development* using *small representative examples to large scale projects*. The

E. Sekerinski and L. Ribeiro (Eds.): FMTea 2024, LNCS 14939, pp. 106–123, 2024.
https://doi.org/10.1007/978-3-031-71379-8_7

necessity for including formal methods in computer science and engineering curricula is described and argued in more detail in the recent article [7].

The approach is presented using the Sireum [32] framework developed at KSU – a collection of model/language processing infrastructure, industry relevant model-based technologies, automated model and code contract verification, developer-friendly manual proof approaches, property-based testing, and assurance case tools. Sireum development has been funded through a variety of US Department of Defense projects in collaboration with industrial research partners at Collins Aerospace and Galois, who have excellent reputations in practical use of formal methods. A corner stone of Sireum is Slang (the "*Sireum Lang*uage") – a safety-critical subset of the Scala programming language [28]. Slang is accompanied by Logika, an advanced verification tool for code-level automated contract-based verification using symbolic execution, which also integrates developer-friendly manual proofs and property-based testing [29].

Interestingly, the first versions of Slang and Logika were developed by the third author (Robby) in 2015 as tools for teaching truth tables, natural deduction, and Hoare logic to undergraduates at KSU. This initial version only treated a very simple imperative programs in single script files. However, Logika's intuitive presentation, IDE-based explanations of reasoning steps, and success with students inspired a much larger vision for rigorous engineering, eventually leading to the long-term industry funding and collaborations described above. These collaborations included building High Assurance Modeling and Rapid engineering framework (HAMR) [17] – a model-based development framework for the Architecture Analysis and Design Language (AADL) and SysMLv2 modeling languages that supports multi-platform development, including, e.g., development of verified applications [5] that run on the verified seL4 microkernel [20].

Throughout this history, various versions of Slang and Logika have been used to teach over 1000 undergraduate students at KSU. The KSU undergraduate course utilizes several novel pedagogical features of Logika including built-in notations for natural deduction proofs and step-by-step Hoare-logic proofs of program correctness. Demonstrating the ability for our material to engage other faculty, Julie Thorton, a teaching professor at KSU now runs the undergraduate course and has developed an online textbook for the course [31]. Subsequently, the first author (Hallerstede), developed course material for Slang, Logika, and HAMR for both undergraduate and graduate courses at AU. These courses have been co-taught with multiple AU faculty.

From this very broad collection of pedagogical material, in this paper, we highlight the strategy that Hallerstede has developed for a graduate course called *Software Correctness*. Thes specific goals of this paper are to:

- provide an overview of Slang and Logika that communicates both their pedagogical value as well as their use in developing real systems,
- describe the pedagogical strategy and course outline of the Software Correctness course,
- illustrate selected aspects of how Logika is used in the course,
- describe how we have connected teaching of formal methods across multiple courses at both AU and KSU, leading to interesting applications for class projects including medical devices and nuclear power plant shutdown systems.

- give references to publicly available pedagogical material that enable other instructions to easily incorporate this content into their own teaching.

2 Slang

Slang is a high-integrity subset of Scala designed from the outset to support developer-friendly integrated contracts and proof notations using SMT-based symbolic execution. Slang retains some of the expressive higher-level features of Scala (classes, traits, higher-order functions) while restricting them to a form that enables more effective verification (e.g., aliasing restrictions). A subset of Slang (called "Slang Embedded") is further restricted to constructs that can be translated to C and Rust appropriate for high assurance embedded systems (without garbage collection runtime). For additional assurance, Slang-generated C code can be compiled using the CompCert verified C compiler [24]. For a detailed overview of Slang features and design rationale, see [28].

Why Slang?: Students inevitably inquire about the choice of Slang as the programming language for the course. We explain that Slang is one of several languages including SPARK Ada [19] and Dafny [23] that were designed to support tool-based reasoning about programs but still be feature-rich enough to support development of real systems. We note that we previously developed contract verification and information flow analysis tools for SPARK [4], and that SPARK inspired the design of Slang. We point to Dafny as a framework that is similar to Slang/Logika in that it is a higher-level language designed for verification from which code in lower-level languages like C can be generated, and we point out applications of Dafny at Amazon. We also note the broader trend toward languages like Rust whose features are designed to eliminate categories of common errors while also clarifying reasoning.

Scalable: While in many cases, examples are presented using simple Slang scripts (based on Scala scripts), we note that Slang supports development of very large programs, e.g., it is used to code over 250K SLOC in the Sireum framework itself, including the Logika verification engine and Slang language processing tools. Sireum includes an IDE-integrated parallel build tool that additionally provides incremental analysis and verification of Slang code/contract, even in the context of large system developments.

Interesting Applications: Slang code can be integrated with Scala and Java in a variety of implementation strategies. Using the ScalaJS transpiler, Slang can also be translated to Javascript. While Slang has been used to program many interesting applications, we typically emphasize to students its use in the KSU HAMR [17] model-driven development framework. Given an AADL [30] or SysMLv2 component-based system architecture model, HAMR generates AADL runtime services [18] in Slang Embedded that can be deployed in various platforms, including the seL4 verified micro-kernel (via C) with formal evidence that architectural constraints are preserved, helping to guarantee safe/secure inter-component spatial and temporal separations [5]. The smallest HAMR AADL-based system – a building temperature control system, can be deployed on a STM32 board with only 192Kb SRAM.

Industrial Research Projects: We motivate the students by pointing out that industry development of safety-critical systems is increasingly emphasizing the types of approaches presented in the class. We describe how the development of Slang, Logika, and HAMR were funded by the US Defense Advanced Research Projects Agency (DARPA), US Army, US Airforce Research Labs, in collaboration with engineers at Collins Aerospace and Galois. On the recent DARPA CASE program, HAMR was used by Collins engineers to build an experimental mission control subsystem running on seL4 for the Boeing CH-47 Chinook helicopter platform [5]. While some of this development emphasized C, we discuss how the HAMR code generation and runtime libraries are implemented in Slang and verified in part with Logika, and how HAMR code generation for Slang and Logika is guiding our new capabilities for generating Rust with integrated contracts for verification with the Verus verification tool [22].

Language Independent Concepts: We try to explain basic principles of specification, testing, and verification in a language independent way. This is accomplished in part by having students use code comments in early exercises to notate what they believe are the facts that hold at particular program points. We also describe how they can apply reasoning principles even when a language does not have a supporting verifier, and we illustrate contracts and verification in other languages including Java (JML), SPARK, and C# (Microsoft Code Contracts). This approach to teaching formal methods is very much in line with the staged approach outlined in [9] progressing from informal to increasingly rigorous and formal reasoning.

Connections to other Courses: At both AU and KSU, Slang and Logika are used in other courses, for which there are opportunities to apply formal methods concepts in other contexts. For example, a KSU graduate course in High Assurance Systems emphasizes HAMR for model-based development of systems written in Slang [16] and verified with Logika. Student course projects include using HAMR to develop the software for medical devices including an infant incubator, a Patient Controlled Analgesic (PCA) Pump, and a subsystem of the Galois HARDENS [14] nuclear reactor trip system (e.g., complete artifacts for the HAMR/Slang/Logika implementation of the HARDENS subsystem can be found at [15]). Teaching resources available include lecture slides, lecture recordings, both written and video tutorials for applying HAMR for the AADL modeling language and Slang [16]. These projects emphasize using Logika's contract specification and verification as well as automated property-based unit and system testing.

At AU, a new software engineering curriculum in computer engineering has been established [11] where Logika is used in the third semester BSc course Programming and Modelling. The new curriculum exposes the students to *formal methods thinking* [9] from their first semester Introduction to Programming onwards. The third semester course Programming and Modelling introduces the students for the first time to tool-based formal methods based on Slang, Logika and HAMR as well as VDM and INTO-CPS described elsewhere [21]. The AU MSc course Software Correctness is described in more detail in Sects. 4 and 5.

3   Logika

Slang's contract language supports assertions, method pre/post-conditions, data type invariants, and global invariants for global states. Verification of code conformance to contracts is performed compositionally and employs multiple back-end solvers in parallel, including Alt-Ergo [8], CVC4 [3], CVC5 [2], and Z3 [26]. Logika uses a forward

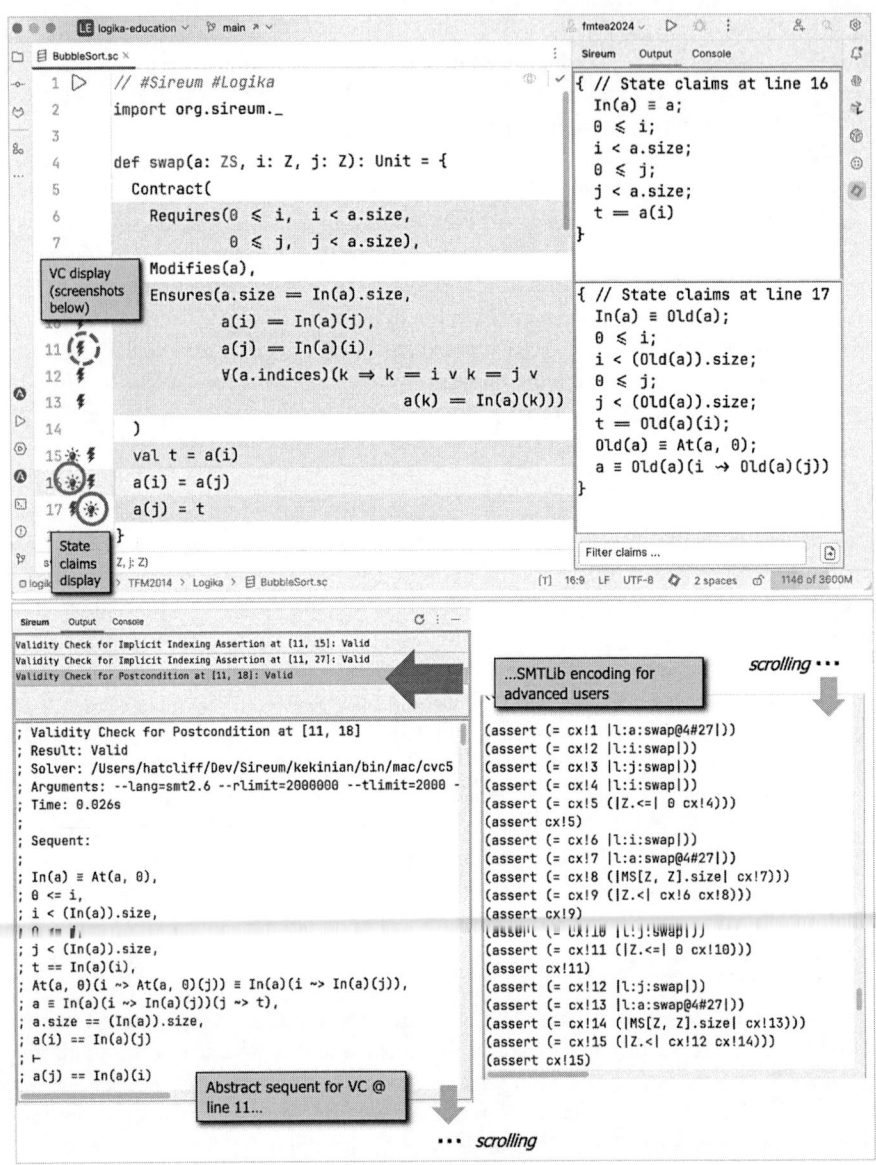

Fig. 1. Logika IVE Features

verification approach based on symbolic execution instead of a backward approach based on weakest pre-condition construction. Based on two decades of experience of implementing symbolic execution tools for both Java and SPARK, we believe that the symbolic execution approach produces diagnostic information about verification steps that is much easier to understand and also allows a more intuitive summary of the underlying verification algorithm for students. The scalability of Logika is complemented by using incremental, focused, and parallel (distributable) verification algorithms. For situations where automated solvers cannot provide full verification, Slang includes an extensible proof language (providing, e.g., term rewriting) directly integrated with the programming language that Logika checks. Verification results, developer feedback on verification status, and contract/proof editing are supported in the Sireum Integrated Verification Environment (IVE) – a customization of the popular IntelliJ IDE. With the IVE, we are able to teach testing concepts with conventional unit libraries for Scala (e.g., ScalaTest) as well as explain testing coverage concepts using IntelliJ's built in coverage facility. With a full build system, we are able to seamlessly transition from teaching basic concepts using small script files to larger multi-file projects.[1]

Figure 1 shows an example Slang method in the Sireum IVE that swaps the values at index positions i and j in sequence a. The swap method has a contract pre-condition (Requires) that constrains the index values to be in range, a post-condition (Ensures) that captures the effect of swapping, and frame condition (Modifies) indicating that only the sequence a is modified. As the developer types, Logika continuously runs in the background to check for possible run-time exceptions, assertion violations, and to check that the code conforms to declared contracts.

Significant engineering efforts have been devoted to displaying verification results directly in terms of program features that the developer can recognize instead of lower level representations such as information flowing to/from the SMT solver. The engineering involves mapping Logika's three-address code intermediate representation and internal logic variables back to program expressions and variables, and also maintaining mappings between program artifacts and SMT-LIB encodings. One of Logika's most distinguishing features (which is especially relevant for education) is to make this verification information available to developers at each relevant point in the code via clickable annotations in the left margin of the editor. There are two types of information: the lightbulb icons ☀ display *facts* roughly corresponding to statement level pre/post-conditions and lightning bolt icons ⚡ display sequents representing verification conditions that are encoded as calls to underlying SMT solvers.

As Logika works, it collects facts that it discovers by moving forward step-by-step through the code. Some of the accumulated facts are immediately apparent from the structure of each program statement (we will refer to these as *immediate facts*). Others are the result of deductions that it has made by calling the underlying SMT solvers (we will refer to those as *deduced facts*). Logika can display the facts that it has accumulated at any point in the program (via the bulb icons) to provide valuable hints about how to systematically reason about the program. These inferred facts are computed based on Logika's internal symbolic execution path conditions that must hold at the particular program points. The top right side of Fig. 1 shows the facts displayed by clicking the bulb annotations at lines 16 and 17. The facts at line 16 reflect constraints associated with the precondition, and the last line represents the outcome of assigning a(i) to t at line 15. The facts at line 17 reflects the execution of line 16 and the Old(..)

---

[1] See https://doc.sireum.org/venues/presentations/logika/tccoe22/ for a video of a 25-minute technical talk and demonstration of Logika's IVE user interface and server-based checking architecture.

annotation is used to distinguish between the values of a in the pre-state and post-state. As will be explained later in the paper, students are first taught the principles of accumulating facts and they record them in comments in their code. They can check their intuition using the bulb information.

The bolt annotations indicate the points at which Logika makes automated deductions that require interactions with its underlying SMT solvers. Logika calls these interactions *summonings* because the power of SMT solving is being "summoned" to make a deduction that cannot be carried out using simple syntactic manipulation of the current facts. Clicking on the lightning bolt shows the details of the summoning. Generally, summonings occur for each line of a post-condition, for assertions, invariants, for checking the pre-condition of a called method, and for code branches to determine the feasible path(s) along which verification should proceed. The bottom of Fig. 1 shows selected aspects of clicking on the post-condition clause at line 11. The left side shows that there are three summonings at this line: two for implicit requirements that i and j are in bounds, and the last corresponding to the explicit claim about the new value at position j. The user clicks on a particular summoning description (e.g., as indicated by the red arrow in the figure) to see the information for that summoning. The most useful information to students and developers is a sequent that abstractly represents a verification condition. The antecedents of the sequent represent the accumulated facts and the consequent represents the condition to be proved (e.g., the new value at position j has the appropriate value). Student are first taught the principles of making deductions from program facts (which they can specify and verify using Logika's Deduce construct). They then learn to check their expectations with the bolt display.

Scrolling down in the output shows the encoding of the sequent as an SMT query in terms of Logika's lower level variable/program representation (illustrated with a fragment in the bottom right of Fig. 1). We explain the concepts of this encoding in graduate classes. For example, students have an assignment in which they: (a) describe the purpose of each line in SMT encoding, and (b) they manually construct simple SMT queries for verification following Logika's encoding. Other information in the Output pane in the bottom left of the figure indicates the purpose of the summoning (e.g., a Validity Check for a post-condition) and the summoning result (Valid, Invalid, Don't Know, or Time Out). Other expert-level diagnostic information is also provided such as the command-line options used with the SMT solvers, and SMT time required to solve the query, etc. The SMT query can be exported to a temporary file via a click of a button and edited with syntax highlighting support in the IDE. A key shortcut action is provided to dispatch the (edited) query with the listed specific solver configuration to confirm Logika's result (or to investigate the edited query).

As Logika works, the incremental (parallel) appearance of bolts and bulbs in the margin also indicates Logika's *progress* through the code and contracts as well as the *coverage* achieved (infeasible paths/dead code will contain no annotations).

4   The Software Correctness Course

We teach an elective MSc course on software correctness at Aarhus University (AU). The MSc course uses Slang, supported by Logika within the Sireum integrated veri-

Week 1: Introduction The purpose of the course is explained. Taking the view that software development is an engineering activity, we argue how reasoning about software contributes to its correctness. The use of Sireum, Logika and Slang is motivated, and a small example is considered to allow the students to become slowly familiar with Sireum.

Week 2: Tracing Facts Beginning with how values are traced through programs, with which the students are very familiar, it is motivated how facts, as generalizations of values, are traced through programs. At first only sequences of assignments are considered, and reasoning is practiced based on axiomatic semantics, interpreting programs as predicates and symbolic execution. By the end of this week the students are comfortable with reasoning about simple programs without executing them, using calculation and proof.

Week 3: Conditionals The students learn how control flow is used to trace facts though programs containing if-statements. As the preceding week, the different ways of reasoning are discussed. This approach is also followed in the coming weeks. Compared to the programs of Week 2, the formulas considered become larger and most students are not accustomed to this. It is practiced, how to read and manipulate such formulas. This skill is an essential prerequisite for the coming weeks!

Week 4: Contracts (Test) The use of contracts is motivated discussing the obligations of caller and callee. After the usual terminology and the contract syntax have been introduced, it is discussed how contracts support testing. Equivalence partitioning and boundary value analysis are applied to contracts, and symbolic execution is used to derive test cases for contracts accompanied by an implementation. The students see how the methodology benefits their software development skills while being prepared to carry out verification proofs.

Week 5: Contracts (Proof) Compositional proofs of Slang programs are considered. The students make extensive use of pre- and post-conditions, and learn how come up with these when verifying larger programs.

Week 6: Loops and Recursion The interrelated concepts of induction, recursion and iteration are discussed. Facts are traced through recursive functions and while loops, leading to the concept of (inductive) invariant. Termination and the use of variants are discussed.

Week 7: Unfolding and Fixpoints In order to be able to discuss bounded analysis of loops and recursion, unfolding and fixpoints are discussed. At this stage the students must be comfortable with handling large formulas. The main reason for moving slowly in the beginning of the course is to prepare the students for this. Logika provides specific features to support reasoning by unfolding, permitting the students to check the result of "manual" unfolding against Logika's.

Week 8: Loops and Recursion Testing In Week 6 verification of loops and recursive functions by proof has been discussed. Based on unfolding introduced in Week 7, bounded testing of loops and recursive functions and its relationship to the approach by proof is discussed. Symbolic execution is used to derive test cases. Because of the unfolding this is just the same as in Week 4.

Week 9: Sequences and Arrays Sequences (that can also be used like arrays in Slang) are introduced. More complex contracts and programs are verified making use of formulas making extensive use of quantification. The modification of formulas with quantifiers is challenging. So, the students must be confident about the verification methodology and the use of Logika at this stage.

Weeks 10–11: Verification Examples and Practice Various examples with sequences and structured data-types are considered and more intricate algorithms verified. In order to succeed with this, the students practice how annotate their programs with facts and invariants starting from contracts, and how to use programs as proofs. The students get some familiarity with verification and proof methodology.

Fig. 2. Software Correctness Course Units

fication environment (see Sects. 2 and 3). The course material is developed collaboratively at the two sites. In 2018, we introduced Slang and Logika as a component in an advanced course on programming at AU@. Until 2022, we gained experience with teaching reasoning and verification skills in a programming course and developed the concept for the software correctness course. In 2023, the course has been established at AU@. Before introducing Slang and Logika, we tried theorem provers like Isabelle [27] or Coq [6] that permit some form of functional programming. The students did not see this as part of their program development activities. By contrast, Slang is a programming language and Logika appears like a common IDE@. This turned out to be a major factor for the students to accept their use in program development. About 20 students participate each semester. The teaching takes place 4 h a week as a mixture of (short) frontal lectures interspersed with exercise sessions for smaller problems and tutorial-style sessions for larger problems. Grades are given according to the Danish 7-point grading scale[2] in an oral exam where the attained grade average is approx. 7, in words "good". See Fig. 2 for an overview of the course units. In the first third of the course,

---

[2] See https://ufm.dk/en/education/the-danish-education-system/grading-system/.

basic methodology for reasoning is introduced, in the second, more advanced methodology, and in the third, all is applied to more complex problems. A similar course (using Dafny [23]) has been set up for fourth year undergraduate students at the University of New South Wales [25].

The course emphasizes that programming should be considered an engineering activity: in particular, a software engineer states when software should be considered as functioning correctly and ensures that this is achievedFunctional and imperative programming and corresponding verification techniques are treated conjointly. Reasoning about programs and verification are included as sound programming methodology. To make sure students see immediate benefit of the taught material, motivation is provided in terms of an improvement of:

- programming skills (programming is thinking [13]),
- testing skills (writing meaningful tests, e.g., [1]),
- documentation skills (stating informal contracts and properties), and
- reasoning skills (convincing oneself and others of the correctness of a program).

In addition, the students learn how to use a formal verification tool that permits them to attain certainty about correctness and to master larger and more complex problems. The terminology uses concepts closer to the programmer's view than of that of a logician's and this perspective is supported by Slang and Logika. All of this does not mean, however, that formal methods are not taught or played down. Instead formal methods concepts are motivated as an integral part of programming. This motivation is discussed in the beginning of the course with the students avoiding to devalue their programming skills. We explain that we regard programming as an engineering activity with the objective to *construct correct software*. The students practice the acquired skills by way of a Scala programming project (an implementation of a small graphics application). 10 h during the course are reserved to discuss these topics with respect to their projects. The students reflect on their project and the methods employed in a report that they submit at the end of the course, e.g.: "The challenges we encountered along the way provided valuable insights into the intricacies of ensuring code reliability. Through this process, we came to appreciate the significance of formal verification in testing and guaranteeing the integrity of our software."

5   Teaching Approach in the Software Correctness Course

In order not to distract from the verification methodology, a series of simple examples is used. It is necessary to explain this to the students and make explicit to them what they learn, in particular, early in the course. Gradually, more complex examples are used in small supervised exercise sessions in class. The two main ways to frustrate students with the material appear to be either to suggest that only trivial problems can be addressed or to be too ambitious and risk that the students shipwreck. For this reason, short lectures are interspersed with exercises sessions and later tutorial-style sessions as the examples grow in size and complexity. There are no home assignments except for the programming projects. This approach has been chosen, so that the students are not left alone when carrying out proofs. They can always ask the teacher and are encouraged to do so. Most of the students have a background in software engineering but not in mathematics, so they need some support when they get stuck while reasoning about Slang code.

In the MSc study, the student cohorts are less coherent than in the BSc study because they have different backgrounds from prior degrees. The only prerequisites are a BSc degree in computing and discrete mathematics. Of course, the local computer engineering BSc students have been introduced in formal methods concepts, Slang and Logika, which benefits all students during class sessions as they are encouraged to work in groups and help each other. The teaching tries to accommodate for this by moving from more familiar programming-related topics to verification-related topics. We begin by *tracing values* through programs and, subsequently, generalize this to *tracing facts*. We use the term *fact* instead of predicate to keep the language simple and familiar.

During the course, we mix informal and formal reasoning. Typically, we start informally and then formalize more and more. While reasoning is entirely informal, usually, some facts are stated formally. E.g., that a *variable* z has the *value* 4 is stated as $z \mathrel{==} 4$. In the beginning of the course, we trace the values of variables with facts stated purely as equations. This is easy to relate to the kind of information one could observe in a debugger. We document the reasoning in the program by adding assertions and observe with the students that the informal reasoning improves our understanding of a program. But it has some shortcomings, e.g.: the approach is not very systematic, so we are not sure whether it is sufficient; for larger programs it would become difficult to master all the details. We can thus see the need for more rigor and more automation without disqualifying informal arguments but discussing its limitations. If it is all we have, then it is the best we can do. And, it is beneficial. It results in improved understanding and documentation improve – a point that is made repeatedly throughout the lectures as this effect is observed. Projecting the required reasoning to 100, 1000, ..., 100000 or more lines of Slang code, the need for automation becomes more apparent, independently of whether we talk of proof or testing. Of course, automation is only possible if the methods are rigorous and systematic. To support automation being rigorous should be being formal. The students are used to formality in terms of programming languages. Automated reasoning ensures that all details are kept consistent and in synchrony with Slang programs. This is another aspect that is very familiar to the students.

In the following three subsections, we give a concrete impression how we proceed in the course. Due to space limitations, it is not possible to go into more detail, but we point out some aspects that we believe to be crucial to the success of the course. Figure 2 shows some complementary remarks about other parts and aspects of the course. In Subsect. 5.1, we show how we take the students carefully from thinking about programs in terms of concrete executions with values to abstract executions tracing facts. The initial slow progression is important to support the students understanding. Subsection 5.2 briefly describes how contracts are introduced and used for reasoning. Once contracts are available, recursions and loops are discussed in the course. Finally, Subsect. 5.3 is intended to indicate the level of more intricate examples covered in the last third of the course.

5.1   Execution as Calculation with Values and Facts

A common way to consider a program is to trace its execution following the values that variables take at different times. We do this when debugging programs, where we predict the values of variables and observe deviations. This means, we calculate expected values and compare them to those produced by program execution. Debugging requires that we reason about programs. We can follow a similar approach directly *without executing a program*, stating expected values by asserting them and compare them to those produced by calculation.

```
val m: Z = 3
// deduce m == 3
val n: Z = 5
// deduce m == 3 & n == 5
val z: Z = m + n
// deduce m == 3 & n == 5 & z == 8
val y: Z = z - n
// deduce m == 3 & n == 5 & z == 8 & y == 3
val x: Z = z - y
// deduce m == 3 & n == 5 & z == 8 & y == 3 & x == 5
assert(x == 5 & y == 3)
```

Starting from concrete values, we calculate the values of the variables during the execution of the program. This is easy! We have shown that the program is correct for the provided values. The calculation yields y == 3 & x == 5. It confirms the expected values stated in the final assertion. We informally document the values that we "know" by writing // deduce .... This is better than using assert because it does not affect the execution. We use the final assert as we have no other means yet to learn whether x == 5 & y == 3 holds at the end or not.

What if we do not know the initial values of variables m and n? In Slang we can express this by writing:

```
val m: Z = randomInt()
val n: Z = randomInt()
```

Function randomInt() specifies that an arbitrary integer value is chosen. The rest of the program is unchanged:

```
val z: Z = m + n
val y: Z = z - n
val x: Z = z - y
assert(x == n & y == m)
```

The method for tracing values by calculation does not work if the specific values of the variables are not known. It is also not enough to limit deductions to specific variable values. We need to trace more general kinds of *facts* that *constrain* possible values of variables. A specific variable value is just a special kind of fact that constrains a variable to one value. The calculation uses equalities and algebraic laws. This is not more difficult than the calculation with values.

```
val m: Z = randomInt(); val n: Z = randomInt()
val z: Z = m + n
// deduce z == m + n   (consequence of assignment)
```

```
val y: Z = z - n
// deduce z == m + n   (old fact)
// deduce y == z - n   (consequence of assignment)
// deduce y == m       (proof by algebra)
//                     (y == z - n == (m + n) - n == m)
val x: Z = z - y
// deduce z == m + n   (old fact)
// deduce y == m       (old fact)
// deduce x == z - y   (consequence of assignment)
// deduce x == n       (proof by algebra)
//                     (x == z - y == (m + n) - m == n)
assert(x == n & y == m)
```

Tracing facts through the program, the final assertion can be proved to hold. This step is done carefully in the course. Switching from tracing values to tracing facts requires the students to abstract execution. The program becomes a formal (mathematical) object. In parallel, symbolic execution of (these) programs is considered in the course to reinforce this idea.

Logika can verify all the deductions in the program. In fact, Logika can verify this without any additional information added. Initially, the students can add deductions and check whether Logika confirms them to hold. Later in the course, examples are treated where Logika needs to be supplied with deductions (and invariants) in order to succeed proving.

Restating the program with deductions in Logika, a preamble is required to activate Logika and provide the reasoning commands, such as, Deduce.

```
// #Sireum #Logika
import org.sireum._
```

To make the transition to formal verification in Logika as easy as possible, the informal notation has been chosen to resemble the formal notation. However, in intermediate stages natural language text can also be written as facts in informal deductions. This step is trivial.

```
val m: Z = randomInt(); val n: Z = randomInt()
val z: Z = m + n
Deduce(|- (z == m + n))
val y: Z = z - n
Deduce(|- (z == m + n), |- (y == z - n), |- (y == m))
val x: Z = z - y
Deduce(|- (z == m + n), |- (y == m), |- (x == z - y), |- (x == n))
assert(x == n & y == m)
```

For the students there are two benefits. Firstly, they get the opportunity to sketch suitable deductions informally. The corresponding facts might be syntactically wrong or partly formal giving a place to keep hold of their thoughts. Secondly, informal deduction can also be used without a tool and with other languages. In the introduction of the course, we emphasize that reasoning is possible in any programming language with or without tool support. We keep this theme going throughout the course to ensure relevance beyond Slang and Logika.

Note that in the preceding session, we have carried out a proof that we understand. Logika uses SMT solvers behind the scene. But how Logika really works, does not matter anymore as long as the students understand what it does. Logika and the Sireum IVE permit us to abstract from what theorem provers do and cast these concepts in programming terms.

With this very small set of concepts, we are able to explain the methodology starting with the informal statement of facts, formalizing them, and using Logika to verify those facts. Taking mutable variables into account is only a small step now. We label the mutable variables according to the order in which they are assigned. We refer to variables v labelled by n by means of the expression At (v, n). For the last assignment (with the largest label) we let At (v, n) ==v. Finally, we write in the comment behind each assignment the fact we deduce from it.

```
val m: Z = randomInt(); val n: Z = randomInt()
var x: Z = m      // deduce At(x, 0) == m
var y: Z = n      // deduce At(y, 0) == n
x = x + y         // deduce At(x, 1) == At(x, 0) + At(y, 0)
y = x - y         // deduce y == At(x, 1) - At(y, 0)
x = x - y         // deduce x == At(x, 1) - y
assert(x == n & y == m)
```

To motivate the use of the At notation, prior to stating the facts above, we discuss why mutable variables need to be treated differently from immutable variables. The solution as sketched above is to represent the program with mutable variables as if it was composed of assignments of immutable variables. Once this point has been made, the facts on the right-hand side become obvious. Being given the informal fact statements above, the students can compose a proof by combining them with the proof as described before. Figure 3 shows the deduced facts from example as they are inspected in the Logika IVE@.

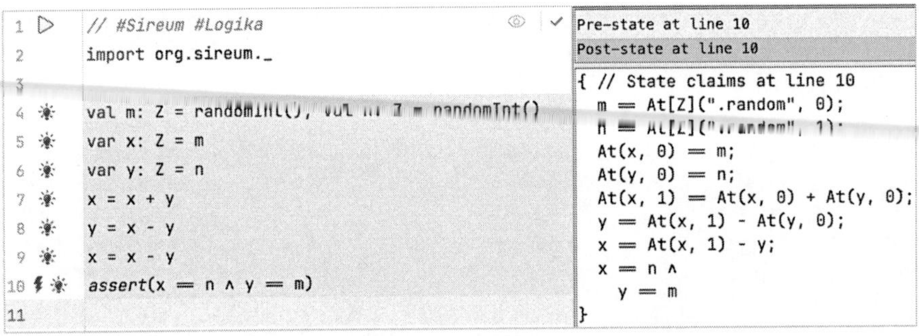

Fig. 3. Inspecting facts pertaining to the program in the Logika IVE

5.2   Contracts, Recursion, and Loops

Contracts are introduced following the same narrative as before, beginning informally, motivating their use for composition verification, and providing intuition underlying the concept. There is nothing difficult about contracts and they are useful for programming and testing. Discussing proof with contracts, however, shows the impact of compositionality concerning proof effort and separation of concerns. Informal contract specifications can be regarded as documentation which may contribute to common documentation frameworks such as Javadoc or Doxygen. This is not to say that these frameworks provide contracts but that being able to write contracts makes for better documentation in these frameworks.

```
def swap(a: ZS, i:Z, j: Z) : Unit = {
  // contract
  //   requires indices i and j are in a.indices
  //   modifies a
  //   ensures values of a at i and j are swapped
  val t: Z = a(i)
  a(i) = a(j)
  a(j) = t
}
```

The contract terminology is coined by simple sentences. (Function swap with the formal contract is shown in Fig. 1.) Using contracts, the discussion of verification of recursive functions is not much different from the non-recursive case because it is no longer necessary to look inside the body of functions.

At this point we observe that we have used contract-like reasoning before in the Slang scripts discussed until now. We have always asserted a fact that we expected to be true at the end. In the course, we also introduce the Slang assume statement. Using it, we have bracketed Slang scripts in pairs of assume-assert statements. Contracts permit us to remove this crutch and lift the reasoning entirely to Logika. Concerning the verification, we can dispose of the assert statement. It is still useful for testing but as an implementation construct. We do not need it to specify expected behavior.

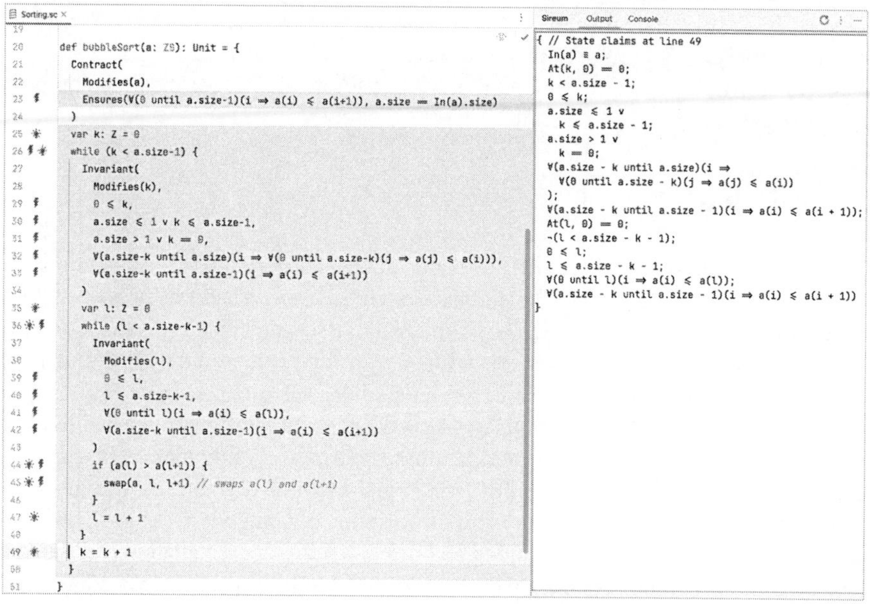

Fig. 4. Logika IVE with array sorting using a nested loop

Based on contracts, we also discuss specification-based testing using equivalence partitioning and boundary value analysis (e.g., [1]). At this stage, we have specification-based testing and implementation-based testing, where we can derive test cases from abstract functional implementations to be used with the concrete implementations. Methodologically, we can prove properties about abstract implementations and derive test cases for concrete implementations. Or, we can use any other mix of verification techniques, including being partly informal. What changes is how the students see software and software development. Reasoning and understanding become independent of execution on a computer.

The approach for verifying Slang programs by testing and proof is gradually extended to include if- and while-statements. Testing of loops is discussed by means of loop-unfolding. It is then possible for the students to explore testing and proof methods for while-loops and assess strengths and limitations.

Testing concepts are supported by the use of the ScalaTest unit testing framework and IntelliJ's test coverage visualizer. To this we add the ScalaCheck automated property-based testing framework. This allows us to connect the predicates used in Logika contracts to the properties expressed in the property-based testing methodology.

5.3   More Intricate Programs

By the end of the course the students verify more intricate programs (as shown in Fig. 4) in tutorial-like sessions where some program fragments are given, e.g., function signatures and informal contracts or parts of contracts. The problems are solved in small groups in class. The lecturer attends the groups, discusses their proofs and helps out when a group gets stuck. At the end, the proof is completed by the lecturer using a projector. Typically, the students add many facts to the program in order to carry out the proofs applying a combination of forward and backward reasoning to reduce the gap between what they know and what they have to prove. The verification and proof methodology, e.g. [10], is discussed in class.

6   Discussion

The Software Correctness Course: We have described our efforts of integrating formal and informal methods into the software engineering curriculums at Kansas State University and Aarhus University. The main objective setting this up was to incorporate the material in such a way that students without a strong background in mathematics or logics, and without a vested interest in formal methods see use in such course. This led to a course design that initially progresses slowly suggesting different ways of thinking about software programs, starting from familiar ways of analyzing them. This gives the students a chance to enter the field of (in-)formal methods without having to master the burden of "theoretical preliminaries" before doing anything interesting. On the way, skills are trained that help in developing test cases, writing documentation and thinking about software programs without executing them. These skills provide immediate value for the students without having to adopt formal methods. Once the students become comfortable with the approach, more complex problems are addressed. The programming (and reasoning) is done in Slang, a dialect of Scala. This makes the whole enterprise look more like programming as there is no need to learn a dedicated modeling language. The Logika IVE seamlessly integrates reasoning with programming permitting the students to verify facts they conjecture and take on large problems. Lecture materials for the course are publicly available [12]. In the obligatory course evaluation the course scores 4.4 (on a scale from 1 to 5). In the free text fields of the questionnaires the students stated, e.g.: "It was nice with a little mini-project to use some of the techniques learned in the course," or "It was really nice to have exercises during the lecture and that [the teacher] walked around to help us if we were struggling with some of the proofs. I really liked that!" But even with the effort made to make the material as relevant as possible for programming, there remains some doubt: "I am not sure if I am going to use what if have learned."

Currently, the course undergoes two main developments: (a) extending the number of examples, in particular, to support self-study, (b) improving the presentation of more advanced verification and proof methodology. This is done relying on discussion and feedback from the students to make sure that they are able to apply directly what they learn. In this respect, it was very helpful to pre-run some of the material in a predecessor course where it was not considered in the oral exam but only in the programming project. The concrete criticism provided by the students concerns course organization but not course content.

One broader benefit of utilizing the Sireum infrastructure for teaching is to connect on "either side" (more basic, more advanced) as described below.

The Basics of Classical Logical and Programming Logic: As mentioned in the introduction, Logika includes dedicated editors for truth tables and natural deduction proofs presented in a syntax oriented to programming. These features are emphasized in the KSU *Logical Foundations of Programming* undergraduate course. The course also emphasizes Logika's manual proof language to help students understand the proof rules associated with Hoare Logika and, more broadly, rules for reasoning about the behavior of imperative programming. An online textbook for the course (providing many illustrations of basic use of Logika for simple examples) can be found here [31].

High Assurance System Development: The Software Correctness course material provides nice preparation for the KSU graduate course on *High Assurance Systems*. This course emphasizes model-based development (using AADL and SysMLv2) of real-time systems using HAMR. The course emphasizes HAMR-based implementations using Slang. Contract-based verification with Logika, along with automated property-based unit testing and system testing are used through course lectures, exercises, and projects. Realistic systems including an infant incubator, PCA Pump, and nuclear reactor shutdown system are used as course projects and are worked "end-to-end" (from requirements to assurance cases). Slides, recorded lectures, tutorials for AADL, HAMR development with Slang, hazard analysis, and broader aspects of safety engineering are available [16].

References

1. Ammann, P., Offutt, J.: Introduction to software testing. Cambridge University Press, 2nd edn. (2016)
2. Barbosa, H., et al.: cvc5: a versatile and Industrial-strength SMT solver. In: TACAS 2022. LNCS, vol. 13243, pp. 415–442. Springer, Cham (2022). https://doi.org/10.1007/978-3-030-99524-9_24
3. Barrett, C., et al.: CVC4. In: Gopalakrishnan, G., Qadeer, S. (eds.) CAV 2011. LNCS, vol. 6806, pp. 171–177. Springer, Heidelberg (2011). https://doi.org/10.1007/978-3-642-22110-1_14
4. Belt, J., Hatcliff, J., Robby, Chalin, P., Hardin, D., Deng, X.: Bakar kiasan: Flexible contract checking for critical systems using symbolic execution. In: NASA Formal Methods, pp. 58–72 (2011). https://doi.org/10.1007/978-3-642-20398-5_6
5. Belt, J., et al.: Model-driven development for the seL4 microkernel using the HAMR framework. J. Syst. Archit. (2022)
6. Bertot, Y., Castéran, P.: Interactive theorem proving and program development. Coq'Art: the calculus of inductive constructions. Springer (2013). https://doi.org/10.1007/978-3-662-07964-5
7. Broy, M., et al.: Does every computer scientist need to know formal methods? Form. Asp. Comput. (2024). https://doi.org/10.1145/3670795
8. Conchon, S., Coquereau, A., Iguernlala, M., Mebsout, A.: Alt-Ergo 2.2. In: SMT Workshop: International Workshop on Satisfiability Modulo Theories (2018)
9. Dongol, B., et al.: On formal methods thinking in computer science education. Form. Asp. Comput. (2024). https://doi.org/10.1145/3670419
10. Gries, D.: The Science of Programming. Springer New York, New York, NY (1981). https://doi.org/10.1007/978-1-4612-5983-1
11. Hallerstede, S., Larsen, P.G., Boudjadar, J., Schultz, C.P.L., Esterle, L.: On the design of a new software engineering curriculum in computer engineering. In: Bruel, J.-M., Capozucca, A., Mazzara, M., Meyer, B., Naumchev, A., Sadovykh, A. (eds.) Frontiers in Software Engineering Education: First International Workshop, FISEE 2019, Villebrumier, France, November 11–13, 2019, Invited Papers, pp. 178–195. Springer International Publishing, Cham (2020). https://doi.org/10.1007/978-3-030-57663-9_12
12. Hallerstede, S., Schultz, C.P.L., Hatcliff, J., Robby: Software correctness course materials. https://github.com/santoslab/software-correctness-course-materials
13. Han, S.: How to write software with mathematical perfection. Quanta Magazine (2022). Interview with Leslie Lamport
14. HARDENS: high assurance rigorous digital engineering for nuclear safety (artifacts repository). https://github.com/GaloisInc/HARDENS
15. HAMR model-based development for the Galois HARDENS reactor trip system (artifacts repository). https://github.com/santoslab/rts-showcase
16. Hatcliff, J.: Course material for high assurance systems. http://s21.highassurance.santoslab.org/lectures.html (2024)
17. Hatcliff, J., Belt, J., Robby, Carpenter, T.: HAMR: An AADL multi-platform code generation toolset. In: Leveraging Applications of Formal Methods, Verification and Validation (ISoLA). LNCS, vol. 13036, pp. 274–295 (2021). https://doi.org/10.1007/978-3-030-89159-6_18
18. Hatcliff, J., Hugues, J., Stewart, D., Wrage, L.: Formalization of the AADL run-time services. In: Margaria, T., Steffen, B. (eds.) Leveraging Applications of Formal Methods, Verification and Validation. Software Engineering: 11th International Symposium, ISoLA 2022, Rhodes, Greece, October 22–30, 2022, Proceedings, Part II, pp. 105–134. Springer Nature Switzerland, Cham (2022). https://doi.org/10.1007/978-3-031-19756-7_7

19. Hoang, D., Moy, Y., Wallenburg, A., Chapman, R.: SPARK 2014 and GNATprove. Int. J. Softw. Tools Technol. Transfer **17**(6) (2015)
20. Klein, G., et al.: seL4: Formal verification of an OS kernel. In: Proceedings of the ACM SIGOPS 22nd Symposium on Operating Systems Principles, pp. 207–220 (2009)
21. Larsen, P.G., et al.: Integrated tool chain for model-based design of cyber-physical systems: The INTO-CPS project. In: 2016 2nd International Workshop on Modelling, Analysis, and Control of Complex CPS, pp. 1–6. IEEE Computer Society (2016).https://doi.org/10.1109/CPSData.2016.7496424
22. Lattuada, A., et al.: Verus: verifying rust programs using linear ghost types. Proc. ACM Program. Lang. **7**(OOPSLA1), 286–315 (2023)
23. Leino, K.R.M.: Program proofs. The MIT Press (2023)
24. Leroy, X., Blazy, S., Kästner, D., Schommer, B., Pister, M., Ferdinand, C.: CompCert-a formally verified optimizing compiler. In: ERTS 2016: Embedded Real Time Software and Systems, 8th European Congress (2016)
25. Morgan, C.: (In-)formal methods: the lost art: a users' manual. In: Liu, Z., Zhang, Z. (eds.) Engineering Trustworthy Software Systems: First International School, SETSS 2014, Chongqing, China, September 8-13, 2014. Tutorial Lectures, pp. 1–79. Springer International Publishing, Cham (2016). https://doi.org/10.1007/978-3-319-29628-9_1
26. de Moura, L., Bjørner, N.: Z3: an efficient SMT Solver. In: Ramakrishnan, C.R., Rehof, J. (eds.) TACAS 2008. LNCS, vol. 4963, pp. 337–340. Springer, Heidelberg (2008). https://doi.org/10.1007/978-3-540-78800-3_24
27. Nipkow, T., Wenzel, M., Paulson, L.C. (eds.): Isabelle/HOL. LNCS, vol. 2283. Springer, Heidelberg (2002). https://doi.org/10.1007/3-540-45949-9
28. Robby, Hatcliff, J.: Slang: The Sireum programming language. In: Leveraging Applications of Formal Methods, Verification and Validation (ISoLA), pp. 253–273. Springer (2021).https://doi.org/10.1007/978-3-030-89159-6_17
29. Robby, Hatcliff, J.: Logika: The Sireum verification framework. In: Formal Methods for Industrial Critical Systems (2024). to appear
30. Society of automotive engineers: Architecture analysis & design language (AADL). Aerospace Standard AS5506 (2004)
31. Thorton, J.: Logical foundations of programming (online textbook for ksu cs 301). https://textbooks.cs.ksu.edu/cis301/index.html
32. Sireum website. https://sireum.org/

Teaching Formal Methods in Application Domains
A Case Study in Computer and Network Security

Achim D. Brucker$^{(\boxtimes)}$ ⓘ and Diego Marmsoler ⓘ

Department of Computer Science, University of Exeter, Exeter, UK
{a.brucker,d.marmsoler}@exeter.ac.uk

Abstract. In this paper, we report on our experience of teaching formal methods as part of an introductory computer and network security module. This module is part of an applied undergraduate computer science degree. As a consequence, we neither can rely on strong theoretical or mathematical foundations of the students, nor can we focus the whole term on applying formal methods in security.

We address these challenges by integrating formal methods into a three-week-long section on security protocols. In these three weeks, we use a holistic approach for teaching the security objectives of security protocols, their analysis of actual implementations using a network sniffer, their formal verification using a model checker (and comparing it to an approach based on interactive theorem proving).

Our approach has been proven successful in teaching (both, in-person and remotely) the benefits of formal methods to numerous students. The students do perform well in the corresponding assessments, and each year we are able to attract students for final year projects (i.e., their B.Sc. thesis) in the area of formal methods.

Keywords: Formal Methods and Security · Protocol Verification · OFMC · Research-led Teaching

1 Introduction

Today, formal methods play an important role in many areas of computer science, with an increasing number of industry sectors integrating them into their development processes. This is not only true in the niche of security-critical or safety-critical systems development (e.g., the use of ACL2 by Green Hills Software [13]). It is also true for cloud and enterprise software companies: For example, Amazon Web Services (AWS) uses their in-house automated reasoning tool ZELKOVA [2] to analyze policies of the AWS Identity and Access Management (IAM) services and the Amazon Simple Storage Service (S3) service. Another example is the use of ASMs by SAP for specifying and analyzing object consistency properties in a distributed system [1].

Despite this success, formal methods are not part of every undergraduate computer science degree: only a finite amount of topics can be covered in any undergraduate degree. Hence, each university offering such a degree needs to make a selection. On the one hand, the authors work at a research-intensive university that offers an applied computer science degree, focusing on software engineering, artificial intelligence, and data science. On the other hand, the authors work at a university that is proud of its *research-led* approach to teaching, and its ability to introduce students to state-of-the art research results. Naturally, the authors of this paper, who are both active researchers in formal methods, are highly motivated to introduce formal methods to their students. As the current environment does not allow for a dedicated "Formal Methods" module (e.g., a module on software verification), the authors opted to integrate an introduction to formal methods into an introductory module on computer and network security.

In this paper, the authors discuss how they teach the application of model checking and its differences to interactive theorem proving as part of a second year computer and network security module. The students (around 200 each year) taking this module do not have any prior exposure to formal methods. At most, they attended a discrete math module that introduced them to the foundations of first-order logic and proof by induction. As part of this computer and network security module, students learn how to formally specify and analyze security protocols. In the lectures, the need for a formal semantics is discussed and the Dolev-Yao [6] attacker model is formally introduced. In the lab sessions, students use a domain-specific model checker, called OFMC [3], to analyze various protocols (e.g., ranging from the famous Needham-Schroeder protocol to simplified versions of TLS). OFMC has been chosen, as it supports a high-level language for specifying security protocols (called Alice and Bob notation) that is also used in the theoretical parts of the lectures. The rest of this module discusses more applied aspects of computer and network security.

The approach presented in this paper has been developed over the course of eight years at two UK universities: The University of Sheffield and the University of Exeter. In this paper, we focus on our experience in Exeter, at which we are offering the module in the form discussed in this paper since the academic year 2019/2020. While the module usually is delivered as synchronous in-person delivery, during the COVID-19 pandemic we also delivered the module successfully as (asynchronous) remote delivery. For the remote delivery, we replaced the lectures by pre-recorded videos and the lab sessions had been taught in flipped-classroom-style, with weekly synchronous drop-in sessions offered online.

The rest of this paper is structured as follows: We start by briefly introducing the environment in which we deliver our teaching integrating formal methods into a generic security module (Sect. 2). Next, we present our approach of integrating formal methods into an introductory security module (Sect. 3) and also, briefly discuss assessment results (Sect. 4). Before we conclude the paper, we briefly contrast our experience in our undergraduate module, with a similar module on the MSc-level, and discuss the wider impact (Sect. 5). Finally, we draw conclusions (Sect. 6).

2   Context

In this section, we briefly introduce the context in which we teach formal methods. Namely, the structure of both the B.Sc. Computer Science as a whole, and of the Computer and Network Security module.

2.1   Structure of the B.Sc. Computer Science

To understand our approach of integrating formal methods into a module on computer security, it is important to understand the context in which this module is taught: "ECM2436 Computer and Network Security" is a mandatory module in the three-year undergraduate program "B.Sc. in Computer Science".

Table 1. Exeter's B.Sc. Computer Science (Optional modules are written in *italic*).

Year 1	ECM1400	Programming	ECM1407	Social and Professional Issues of the Info. Age
	ECM1410	Object-Oriented Programming	ECM1413	Computers and the Internet
	ECM1414	Data Structures and Algorithms	ECM1415	Discrete Mathematics for Computer Science
	ECM1416	Computational Mathematics	COM1011	Fundamentals of Machine Learning
Year 2	ECM2414	Software Development	ECM2418	Computer Languages and Representations
	ECM2419	Database Theory and Design	ECM2426	Network and Computer Security
	ECM2427	Outside the box	ECM2434	Group Software Engineering Project
	ECM2423	*Artificial Intelligence and Applications*	*ECM2425*	*Mobile and Ubiquitous Computing*
	ECM2433	*The C Family*	*ECM1417*	*Web Development*
Year 3	ECM3401	Individual Literature Review and Project		
	ECM3408	*Enterprise Computing*	*ECM3412*	*Nature Inspired Computation*
	ECM3420	*Learning from Data*	*ECM3422*	*Computability and Complexity*
	ECM3423	*Computer Graphics*	*ECM3428*	*Algorithms that Changed the World*
	ECM3446	*High Performance Computing*	*EMP3001*	*Commercial and Industrial Experience*

Table 1 provides an overview of the modules offered in our B.Sc. Computer Science. Except for the B.Sc. Thesis (called "ECM3401 Individual Literature Review and Project"), which is worth 22.5 credits, all taught modules are worth 7.5 credits in the European Credit Transfer and Accumulation System (ECTS). Each taught module is delivered during a 10-week-long teaching period, usually with two lectures and one small-group lab session (of 50min each) per week. Thus, a typical module delivery consists out of 20 taught lectures and 10 lab sessions in total.

Currently, our B.Sc. Computer Science has a cohort of around 180 students. In addition, the module "ECM2426 Computer and Network Security" is also open to students from other programs (e.g., Combined Honors), increasing the number of students for this particular module to around 200 students each year.

As Table 1 shows, our B.Sc. Computer Science is focused on the applied aspects of computer science. There is no dedicated module (neither optional nor mandatory) teaching formal methods. Students interested in theoretical computer science have the choice to take an optional module (ECM3422) in year three. For the students taking our second year Computer and Network Security module, we can only rely on a basic knowledge of first-order logic, set theory, and proof by induction, all of which are taught in the first year as part of the module "ECM1415 Discrete Mathematics for Computer Science". Parallel to the Computer and Network Security module, the students on the B.Sc. Computer Science are taking "ECM2418 Computer Languages and Representations", which is a module teaching Haskell (we will see later, why this is relevant), PROLOG, and finite automata.

2.2 ECM2426 Computer and Network Security

In its delivery, "ECM2426 Computer and Network Security" follows the standard setup for taught undergraduate modules at Exeter: it is delivered as two weekly lectures and one weekly small-group lab session (which is repeated multiple times to serve all students) of 50min each, over 10 weeks in the first term of year two.

As mentioned earlier, this module is *not* a dedicated "Formal Methods for Security" module. According to its module descriptor (this is the document defining what has to be taught in a module and what learning outcomes should be achieved) the intended learning outcomes are:

- Module Specific Skills and Knowledge:
    1. Demonstrate understanding of the concepts, issues, and theories of cryptography and security;
    2. Demonstrate theoretical and practical knowledge of security technologies, tools, and services;
    3. Gain practical experience of developing solutions to networks and computer security challenges.
- Discipline Specific Skills and Knowledge:
    1. Show an awareness of the need for network and computer security;
    2. Demonstrate good design and development skills.
- Personal and Key Transferable / Employment Skills and Knowledge:
    1. Demonstrate practical knowledge of current security methods and tools.

We assess the learning outcomes, firstly, with coursework (contributing 30% to the overall mark), i.e., a practical "take home" assessment for which students have several weeks for completing it. Secondly, students have to sit a closed-book exam of 90 min (contributing 70% of the module mark).

In designing the module curricula, we assumed that the majority of our students will at least start their career in a software development related role (e.g., as web developer or software engineer). This assumption, together with the intended learning outcomes, guided the selection of topics, which can roughly be classified into four areas:

1. *Foundations and Access Control:* Introducing the fundamental concepts of information security (i.e., confidentiality, integrity, and availability) as well as access control as one means of controlling access to information.

2. *Cryptography and Digital Signatures:* Introducing symmetric and asymmetric encryption as well as cryptographic hashes and digital signatures. The discussion of x.509 certificates in web browsers prepares the next part of the module.
3. *Network Security and Security Protocols:* Introducing network security, mostly in the form of discussing protocols for establishing confidential and authentic communication channels. In this part, we cover both, the applied aspects (e.g., sniffing network packages) and the formal modelling and analysis of such protocols.
4. *Software Security:* Introducing secure software development, including software vulnerabilities, secure (defensive) programming, or security testing (for software developers).

Table 2. Curriculum of ECM2426 "Computer and Network Security": The module focuses on four areas of computer and network security: Foundations and Access Control , Cryptography and Digital Signatures , Network Security and Security Protocols (SP) , and Software Security

	Lecture	Lecture	Lab
Week 1	Confidentiality, Integrity, Availability	Access Control	–
Week 2	Symmetric Encryption 1	Symmetric Encryption 2	CIA and Access Control
Week 3	Asymmetric Encryption 1	Asymmetric Encryption 2	Symmetric Encryption
Week 4	Hashing, Digital Signatures & PKIs	Cryptoanalysis	Asymmetric Encryption
Week 5	Privacy vs. Security	Security Protocols 1	Security Protocols
Week 6	Reading Week		
Week 7	Security Protocols 2	Formal Analysis of SPs 1	Formal Analysis of SPs
Week 8	Formal Analysis of SPs 2	Research Spotlight	Formal Analysis of SPs
Week 9	Software Security & Vulnerabilities	Secure Programming	Manual Security Analysis
Week 10	Security Testing 1	Security Testing 2	Static Code Analysis
Week 11	Threat Modelling & Vulnerability Disclosure	Topic Selected by Students	Dyn. Security Testing
Week 12	Exam Preparation 1	Exam Preparation 2	Forensics

Table 2 shows the teaching curriculum in more detail. Note that week 6 of term 1 is "Reading Week", i.e., a week in which students are encouraged to self-study and revise previously taught material. There are no taught classes in Reading Week. Week 12 is the so-called "Revision Week", in which no new content is taught: the focus of this week is on revising taught content, usually in the form of exam preparation.

3   Teaching FM as Part of a Unit on Network Security

In the following, we will focus on the ca. two weeks long "Network Security and Security Protocols" part of the module, in which integrated an introduction into formal modelling and formal analysis.

3.1   Design and Intended Learning Outcomes

When designing this two and a half weeks long part of the Computer and Network Security module, we were guided by four aims:

1. It has to contribute to achieving the intended learning outcomes of the module (recall Sect. 2.2). This also includes security-specific learning outcomes, such as understanding that systems that use strong cryptography can still be insecure, or that composing two secure systems does not necessarily result in a secure composed system.
2. It has to be attractive to both, students interested in theoretical and formal topics *and* students that are more interested in the practical aspects of network security.
3. It should achieve a number of formal methods specific intended learning outcomes that otherwise are not (or hardly) covered by our B.Sc. in Computer Science. They fall into two categories:
   (a) Security-specific learning outcomes, such as the value of model-checking for finding attacks on security protocols.
   (b) Transferable formal methods skills, such as understanding the need and value of a formal semantics of a specification (or programming) language, understanding the concept of undecidability, or the difference between a semi-decision procedure and a decision procedure.
4. It should allow us to refer to our own research, giving student an insight what we, as academics, are doing, when we do not teach. Furthermore, it should motivate students to consider a B.Sc. thesis related to formal methods as well as consider a PhD in this area.

The intended learning outcomes fall into two categories: first, there are security-specific learning outcomes, such as understanding that systems that use strong cryptography can still be insecure or that composing two secure systems does not necessarily result in a secure composed system. Second, there are formal-methods-specific learning outcomes: understanding the need for correct and secure systems and the role formal verification can play in providing these guarantees. Furthermore, students should develop an understanding of the limitations of formal verification approaches. To facilitate the latter, recent research of one of the authors [10] on a verification approach for security protocols using interactive theorem proving is demonstrated to students (but students do not use interactive theorem proving themselves).

For our integration of formal methods, we have selected OFMC [3] as tool for several reasons: most importantly, OFMC supports the Alice and Bob notation as specification language for security protocol. This allows us to teach formal methods thinking without the need to spend a lot of time on mathematical formalisms that our students might struggle with.[1] Second, our group has research

---

[1] Note that the Alice and Bob notation is a formal language, and in fact its semantics is rather involved. However, it is especially designed to be simple and intuitive to use, and it allows for a presentation that can be linked to the Dolev-Yao model.

links to the group developing OFMC, which gives the use of OFMC a "personal touch" that, in our experience, is appreciated by students. Finally, OFMC is implemented in Haskell, and, as our students learn Haskell in a module that runs parallel to the security module, the use of OFMC provides an example of a non-trivial program implemented in Haskell.

3.2   Motivation

We start the block on security protocols by introducing the Needham-Schroeder Public Key protocol [15]. At its core, it is a small protocol consisting out of only three steps (see Fig. 1a) that, assuming an already existing public key infrastructure, allows two agents (usually called Alice and Bob) to establish a mutual authentication property: after the completion of the protocol, only Alice and Bob should know the freshly generated random values (usually called nonces, for "numbers only used once") N_A and N_B. Given the informal correctness argument (Fig. 1a), it is convincing that the protocol achieves its security goal.

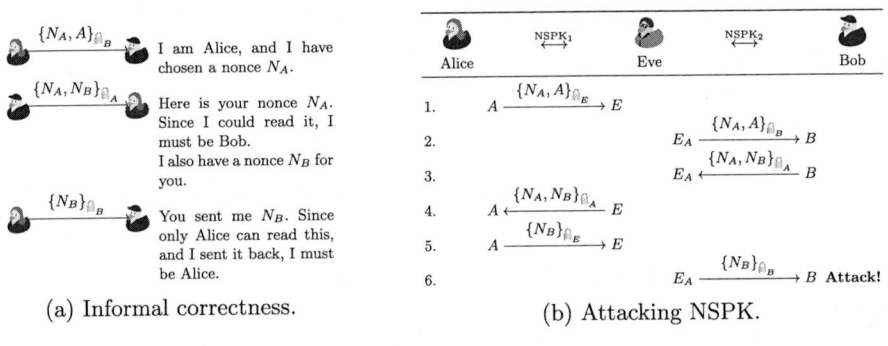

(a) Informal correctness.   (b) Attacking NSPK.

Fig. 1. The Needham-Schroder Public Key (NSPK) Protocol.

This protocol was first published in 1978, and it took until 1995 to discover an attack (see Fig. 1b) [11]. Here, an attacker (called Eve), is able to trick Bob into believing she is Alice. In the second step, Eve, instead of replying directly to Alice, plays unfair, i.e., does not follow the rules: Eve re-uses the nonce in the message from Alice (this re-use of a nonce is violating the rules of the protocols, hence, we call Eve a dishonest agent) to start a run of the NSPK protocol with Bob. Furthermore, in the first message from Eve to Bob, Eve claims to be Alice. Consequently, Bob responds to that message with a new nonce in a message encrypted to Alice (as Eve claimed to be Alice in her first message). Therefore, Eve cannot read (decrypt) this message. Still, this second message from the protocol run between Eve and Bob looks the same as the second message Alice is waiting for in the protocol run that she started with Eve. Hence, when receiving this message, she responds with the third message, encrypted for Eve. When Eve receives this message, she learns the nonce generated by Bob and, consequently, can finish the protocol run with Bob. As a result, Eve was successfully able to convince Bob that she is, indeed, Alice. We then discuss Lowe's fix [11] and, informally, discuss how and why it prevents this particular attack.

During this part, we rely on an intuitive understanding of the notation used (at this stage, we do also not yet distinguish between roles and agents, i.e., the concrete people playing a role). The main intended learning outcomes of this section are: firstly, that even small (distributed) systems are hard to build correctly (and hard to understand); secondly that just because something is encrypted, it does *not* mean it is confidential. And, last but not least, we still do not know whether the protocol is secure against other attacks.

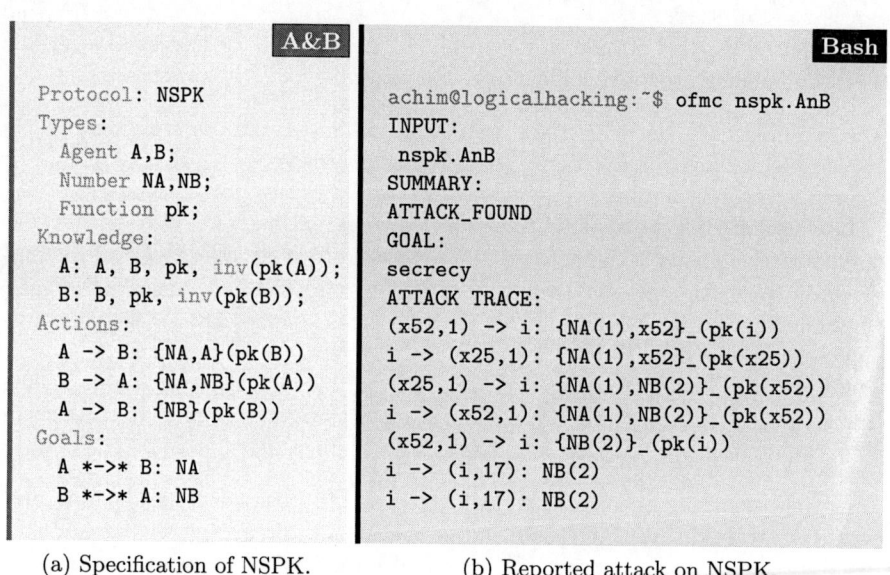

(a) Specification of NSPK.   (b) Reported attack on NSPK.

Fig. 2. Analyzing NSPK with OFMC

To give students a first outlook where we are heading to, we usually end the first lecture on security protocols with a brief demonstration of OFMC [3], analyzing NSPK (see Fig. 2). Here, OFMC "re-discovers" the attack on NSPK in less than a second (Fig. 2b). Note that at this point of the module, we rely on an intuitive understanding of the specification (Fig. 2a) in Alice and Bob notation. The focus is more on showing students that such attacks can be found by automated tools.

3.3 Modelling Security Protocols

In the next two lectures we continue to informally introduce notation for security protocols, by designing a protocol from first principles. We use the same scenario that is also used in the OFMC Tutorial [12]: the iterative design of a key establishment protocol using a trusted third party (usually called a key sever). We start with a very naive protocol using the roles A, B, and the trusted key server s:

1. $A \longrightarrow s : \quad A, B$ An agent playing role A contacts s, sending the identifiers of the two parties wanting a shared session key.

2. $s \longrightarrow A : \quad K_{AB}$ s sends the session key K_{AB} to A.

3. $A \longrightarrow B : \quad K_{AB}, A$ A passes K_{AB} on to B.

This protocol is, of course, not secure, if we assume that an attacker can eavesdrop on all messages sent in a security protocol. We call this our first security assumption.

Furthermore, this simple protocol allows us already to discuss several key aspects of such protocol specifications. For example:

- The differences between roles (denoted by uppercase letters) and agents (denoted by lowercase letters). Agents can play roles.
- Sender/receiver names (e.g., "$A \longrightarrow B$") are not part of the messages.
- Messages are not guaranteed to reach their destination (securely). Recall that the intruder controls the network, and can intercept any message and remember it (even though maybe not decrypt it) and send messages under any sender identity. In essence, this is the worst case of an intruder who controls the entire network.
- We specify here how agents are *supposed* to behave, and the honest agents would stick to that, while the dishonest agents are only bound by their cryptographic abilities, and are summarized as the intruder.

In the remaining two lectures dedicated to this part, we re-design this protocol iteratively, in each iteration fixing the attack discovered in the previous version of the protocol. During the course of refining this protocol, we

- Discover several types of attacks on security protocols, such as replay attacks, person-in-the-middle attacks, or oracle attacks.
- Discover four fundamental security assumptions that, informally, capture the Dolev-Yao intruder model [6].

We end this part of the module by formally introducing the syntax for specifying security protocols.

Many of our students would consider a purely theory-focused module (or a series of lectures for that matter) rather "dry" and not very attractive. To keep also students engaged that are not much interested in the theoretical foundations, we integrate practical demonstrations of attacks into our teaching. For example, we have a small demonstration using a smart light bulb that allows an attacker to replay commands sent to the light bulb. We demonstrate this attack by sniffing the network communication between the mobile app used for controlling the light bulb and its back-end servers using Wireshark (https://www.wireshark.org/), and replaying the observed commands using curl (https://curl.se) on the

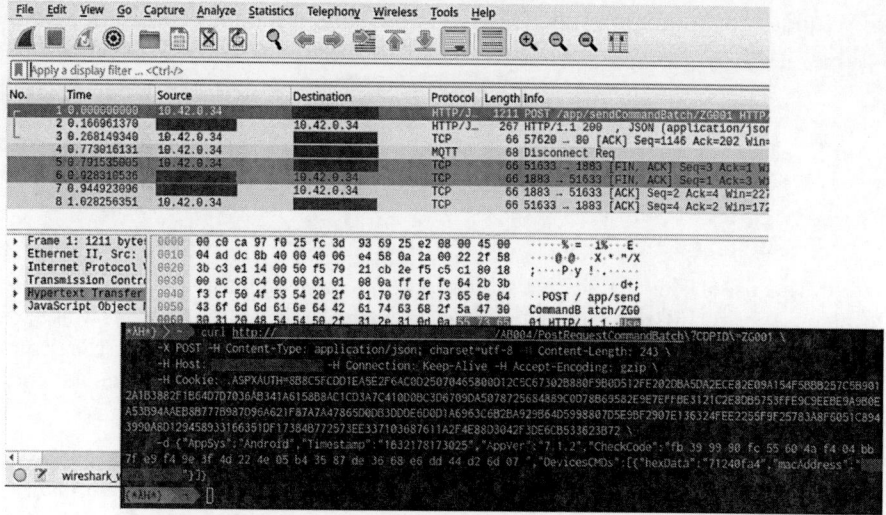

Fig. 3. Analysing network traffic using WireShark and executing a replay-attack using curl.

Linux command line (see Fig. 3). In the corresponding lab, we provide students with a virtual machine that simulates a network with several servers and clients that run insecure protocols (e.g., http, telnet, ftp). In this lab, students learn how to use nmap (https://nmap.org) for exploring the network and Wireshark for analyzing individual protocol sessions, discovering several security flaws.

3.4 Analyzing Security Protocols

After spending one week on introducing security protocols and informally motivating the Dolev-Yao attacker model [6], we now use the following one-and-a-half weeks to motivate the security protocol verification problem and teach our students to understand OFMC, as one example of a semi-decision procedure.

Decidability and Rice's Theorem. We start this part of the lecture by introducing Rice's Theorem, to motivate that the protocol verification problem is undecidable:

Theorem 1 (Rice's Theorem). *Let S be any non-empty, proper subset of the computable functions. Then the verification problem for S (the set of programs P that compute a function in S) is undecidable.*

As our students did not have any prior exposure to the concept of decidability, we take quite some time to motivate it pragmatically: for example, we jointly repeat the analysis of the original Needham-Schroeder protocol (recall Fig. 2) with OFMC and apply Lowe's fix (changing the second message of the protocol to {NA, NB, B}(pk(A))) [11]. For the fixed version, we demonstrate that

OFMC does not terminate (i.e., OFMC implements a semi-decision procedure), iteratively exploring the search space:

```Bash
Open-Source Fixedpoint Model-Checker version 2020
Verified for 1 sessions
Verified for 2 sessions
...
```

For the proof sketch of Rice's Theorem, we use reduction to the Halting Problem. While this requires us to introduce the Halting Problem as well, we believe that this is worthwhile for two reasons: firstly, the proof for the Halting Problem seems to be easier accessible for computer science students and, secondly, it allows discussing decidability from a slightly different angle.

Alice and Bob: Syntax. Next, we formally introduce the Alice and Bob Notation for specifying security protocols. This requires (recall Fig. 2a) revising the concepts of types (which most students should learn in a parallel offered module teaching Haskell), the concept of the initial knowledge of agents, the actions, and the security goals (i.e., secrecy and/or authenticity of message parts). To help students, we also use simple message-sequence-charts for describing the actions. The latter also allows us to discuss how such a protocol can be split into chords for the individual roles. This prepares the discussion of the semantics of such protocol specifications. But before we discuss the semantics, we introduce our attacker model formally.

The Dolev-Yao Attacker. Informally, the Dolev-Yao intruder is often described as an attacker who has full control over the network, i.e., they can read and block all messages sent, send messages on their own, and decrypt messages if they have the required private key (as well as read the content of signed messages). We define the Dolev-Yao intruder formally as a set of rules of a logical calculus, defining the Dolev-Yao Closure:

Definition 1 (Dolev-Yao Closure). *Given a set of terms M, we define $\mathcal{DY}(M)$ as the least closure of M under the following rules:*

$$\frac{}{m \in \mathcal{DY}(M)} \; Axiom \; (m \in M) \qquad \frac{s \in \mathcal{DY}(M)}{t \in \mathcal{DY}(M)} \; Algebra \; (s \approx t)$$

$$\frac{t_1 \in \mathcal{DY}(M) \quad \cdots \quad t_n \in \mathcal{DY}(M)}{f(t_1, \ldots, t_n) \in \mathcal{DY}(M)} \; Composition \; (f \in \Sigma_p) \qquad \frac{\langle m_1, m_2 \rangle \in \mathcal{DY}(M)}{m_i \in \mathcal{DY}(M)} \; Proj_i$$

$$\frac{\{\!|m|\!\}_k \in \mathcal{DY}(M) \quad k \in \mathcal{DY}(M)}{m \in \mathcal{DY}(M)} \; DecSym$$

$$\frac{\{m\}_k \in \mathcal{DY}(M) \quad inv(k) \in \mathcal{DY}(M)}{m \in \mathcal{DY}(M)} \; DecAsym \qquad \frac{\{m\}_{inv(k)} \in \mathcal{DY}(M)}{m \in \mathcal{DY}(M)} \; OpenSig$$

Discussion. One might ask why we do not discuss the syntax and semantics of the Alice and Bob Notation at the same time. This has mostly organizational reasons: With the Dolev-Yao Closure, we conclude the part of the security protocol topic that students are assessed on.

We support the learning of this part with two lab sessions. In the first, we focus on the use of OFMC and a more pragmatic understanding of security protocol specifications. To this end, we ask students to model, analyze, and fix various smaller protocols using OFMC. We also experiment with different initial knowledge sets, for instance, discussing the impact of under/over-approximations on the verification result. In the second, we work with the formal definitions introduced, i.e., we discuss other semi-decidable problems and, most importantly, we use the Dolev-Yao Closure to derive if an attacker, given a set of messages, is able to infer a certain secret (see Sect. 4).

3.5   Research Spotlight

We conclude this part of the module with a lecture that is not part of the assessed content. This is a compromise we make to give students interested in formal aspects of security a "deep-dive", while allowing students that are less interested to focus on the remaining units of the module.

In this part of the lecture, we introduce the free term-algebra and discuss the semantics of the Alice and Bob notation by describing roles as a sequence of events that either send $(\mathrm{snd}(\_))$ or receive $(\mathrm{rcv}(\_))$ messages. For example, role A of the NSPK protocol can be described as:

$$\mathrm{snd}(\{N_A, A\}_{\mathrm{pk}(B)}) \cdot \mathrm{rcv}(\{N_A, N_B\}_{\mathrm{pk}(A)}) \cdot \mathrm{snd}(\{N_B\}_{\mathrm{pk}(B)})$$

This allows us to introduce an operational semantics for roles and sketch how it can be used for discovering the attack on the NSPK protocol.

As mentioned earlier, one reason for selecting OFMC as tool of choice is the strong links our research group has with the group developing OFMC. Actually, we are jointly developing a protocol verification tool, called PSPSP [9,10], based on the interactive theorem prover Isabelle/HOL [17]. As part of the Research Spotlight session, we demonstrate this tool to the students (but students are not expected to use Isabelle or PSPSP themselves). We use this also to show that an interactive tool, using proof by induction (that students know from their first year Discrete Mathematics module) is able to reason over infinite inductively defined (message) sets. Furthermore, some of our students are on a combined B.Sc. Computer Science and Mathematics. These students did use Lean [14] in their first year of their degree program; providing another link between their mathematics and their computer science modules.

4   Assessing the Intended Learning Outcomes

As mentioned earlier, the learning outcomes are assessed with one coursework (homework) and one closed-book exam. Both, the coursework and the final exam are also split into the four main topic areas of the module: Foundations and Access Control, Cryptography and Digital Signatures, Network Security and Security Protocols, and Software Security.

The coursework focuses, in general, on the practical aspects of the intended learning outcomes: each student needs to solve practical exercises on an individualized virtual machine. For the questions testing learning outcomes of the Network Security and Security Protocols, most of the assessment is centered around using tools such as nmap or Wireshark to analyze a network segment with multiple servers and clients. Still, one question usually requires students to use OFMC. For example, students might need to complete an incomplete template of a security protocol provided in Alice and Bob notation. Notably, the coursework makes use of a cloud-based setup that provides each student with a personalized virtual machine that contains all exercises and the software required for solving the exercises. The personalization ensures that each student needs to solve a problem instance that is different to the problems the other students have to solve. For example, each student will get a different protocol template that is generated from a set of base templates by applying well-defined mutation operators. The individual configurations (in the case of security protocol specifications, selecting a template randomly and applying a random selection of carefully designed mutation operators) of the virtual machines are generated during the initial boot process of each virtual machine. Also, during the initial boot process, the configuration of each VM is transferred to a secure storage. A marking tool that we developed uses these configurations to mark the submissions of the students automatically.

In contrast to the coursework, the exam focuses on the foundational and theoretical learning outcomes of the module. Here, we focus on assessing the understanding of the formal specification and analysis of security protocols and the wider understanding of formal methods. For testing the former, we might ask students to apply the rules from the Dolev-Yao Closure (recall Definition 1), to verify if certain messages can be derived by an attacker:

Question 1 Consider the following intruder knowledge

$$M = \Big\{ n_1, n_2, n_3, \mathrm{pk}(a), \mathrm{pk}(b), \mathrm{pk}(c), \mathrm{pk}(i), \mathrm{inv}(\mathrm{pk}(i)),$$
$$m, \{\!|m|\!\}_{hn_1)}, \{\!|m|\!\}_{n_2}, \{h(m)\}_{\mathrm{inv}(\mathrm{pk}(a))},$$
$$\{\{\!|secret|\!\}_{h(m)}\}_{\mathrm{pk}(i)}, \{secret\}_{\mathrm{pk}(b)} \Big\}$$

where g and h are functions (i.e., $g, h \in \Sigma$) and, furthermore, h is a public function (i.e., $h \in \Sigma_P$).
Prove formally that a Dolev-Yao intruder can learn the message "*secret*". Name the rules that you use.   □

To answer this question, students need to be able to construct a derivation tree, similar to natural deduction:

$$\dfrac{\dfrac{\{\!|\,secret\,|\!\}_{h(m)}\}_{\mathrm{pk}(i)} \in \mathcal{DY}(M)}{}\text{Ax.} \qquad \dfrac{}{\mathrm{inv}(\mathrm{pk}(i))}\text{Ax.}}{\{\!|\,secret\,|\!\}_{h(m)} \in \mathcal{DY}(M)}\text{DecAsym} \qquad \dfrac{\dfrac{\{h(m)\}_{\mathrm{inv}(\mathrm{pk}(a))} \in \mathcal{DY}(M)}{}\text{Ax.}}{h(m) \in \mathcal{DY}(M)}\text{OpenSig}$$

$$\dfrac{}{secret \in \mathcal{DY}(M)}\text{DecSym}$$

To assess a broader understanding of formal analysis methods, we ask questions such as:

Question 2 OFMC implements a semi-decision procedure. Briefly explain the consequence of this in case OFMC does find an attack as well as in the case where OFMC does not report an attack. □

Given that our students do not have a deep theoretical knowledge, one could expect that the questions focusing on the application of formal methods score below average. This is *not* the case: the four topic areas carry an equal weight in the exam (i.e., each is worth 25 marks). In the last years, the question on security protocols did actually score slightly better than the average. For example, last year, the four areas achieved the following average marks: Foundations and Access Control: 14.7/25, Cryptography and Digital Signatures: 13.8/25, Network Security and Security Protocols: 14.7/25, and Software Security: 12.7/25. These average marks are representative for the last couple of years. While students usually experience the part on the formal analysis of security protocols as difficult, they seem to prepare very well for it. In contrast, they seem to overrate their abilities in reading program code, which is an important skill for scoring well in the Software Security part of the exam, which usually scores worst.

5 Discussion

In the following we briefly discuss two further aspects: First, transferring the idea of embedding formal methods into a security module to the postgraduate taught level. Second, the wider impact and benefits of teaching formal methods, as research-led topic, to student experience and our research.

5.1 Experience at the Taught Postgraduate Level

Based on our experience in integrating the formal specification and analysis of security protocols, as a showcase of formal methods, into an undergraduate security module, we opted for a similar setup when designing a new MSc-level module. Namely, we integrated a similar two-to-three week long block on formal analysis of security protocols into a module "Fundamentals of Security" that is delivered as part of the M.Sc. in Cyber Security Analytics. The entry requirement for this program is an undergraduate degree equivalent to at least a UK 2:1 Honors degree (i.e., at least an average mark of 60 out of 100 in the UK mark system, where everything above 70 is a "first" and 40 is a "pass") in a STEM undergraduate degree. In particular, the admission teams check that the undergraduate degree includes mathematics and programming modules.

Given the required mathematics prerequisites of the program, we first tried using Isabelle/HOL and a simplified version of Paulson's inductive approach to protocol verification [16], instead of OFMC. This has shown to be too challenging for our students. Firstly, using Isabelle successfully requires more understanding of the underlying theory than OFMC. Secondly, we underestimated how much the knowledge of mathematical concepts amongst our postgraduate students varies. The latter is, most likely, due to the fact that the majority of our postgraduate students did not complete their undergraduate degree at Exeter (and not in the UK, for that matter). As a consequence, we need to serve a cohort of students with a very diverse background. While our undergraduate students do not have a substantial background in theoretical computer science, we do know very well what they learned in their first year. This makes it relatively easy for us to ensure that we use, as much as possible, notation that is already familiar to them. Moreover, it also gives us a good understanding, which aspects we can expect students to know already and which aspects we need to introduce in full detail. In hindsight, this has proven to be an invaluable asset in integrating a formal methods "short course" into an introductory security module.

In contrast, the admissions criteria for our postgraduate program are not fine-grained enough to ensure such common foundations. For example, some students only had statistics in their undergraduate degree, which satisfies the admissions requirements but is not necessarily a good foundation for teaching formal approaches using interactive theorem proving. In the following year, we adopted OFMC also in our postgraduate teaching. While this did reduce the entry barrier for our students, we still felt that we need to spend more time on teaching foundations compared to the undergraduate module.

Currently, we are in the lucky situation that, as part of a larger restructuring of the postgraduate offering of our department, we are able to create a dedicated module on building high-assurance systems. As part of this module, we will be able to spend the time required to provide a more in-depth introduction into formal verification.

5.2   Wider Impact: Individual Literature Review and Project

Based on our integration of formal methods into the mandatory second-year security module, we are able to attract, each year, a few students who choose a formal methods related topic for their Individual Literature Review and Project (i.e., their B.Sc. thesis). As a lot of our research uses Isabelle [17], most of these projects are also using Isabelle.

As our students have no experience in using Isabelle at all and also have only very limited experience in mathematical proofs, we usually design our topics for the Individual Literature Review and Project such that students can start using Isabelle "as a functional programming language": they start with defining data types, simple definitions, and functions similar to what one would do in Haskell. Here, we can benefit from the fact that our students learn Haskell in the "ECM2418 Computer Languages and Representations" module. Furthermore, we usually ensure that the formal specifications developed by our students are executable. This allows the efficient use of Isabelle's "`eval`" command for evaluating ground HOL-terms and also Isabelle's built-in test tool "`quickcheck`". Thus, students that struggle with formal proofs could still complete their project by focusing on a thoroughly tested formal specification: testing the important properties instead of formally proving them.

Over the last couple of years, we had the pleasure of supervising a number of interesting formal methods projects such as the formalization of Blockchains, homomorphic encryption, or the verification of C programs using AutoCorres [8]. At least one of the projects, formalizing neural networks in Isabelle, was directly continued as PhD project [5].

6 Conclusion

Whether formal methods should be a piece in every computer scientist's toolbox, is an ongoing debate [4]. We presented our approach of integrating a "short-course" on formal methods as part of a security module in an otherwise applied B.Sc. Computer Science. Our goal is not to teach formal methods in depth, instead we aim for educating our students in "Formal Methods Thinking," as advocated by Dongol et al. [7].

We did not discuss student feedback in detail. While we did collect feedback in previous years, we only did collect feedback for the overall module as such. Overall, our undergraduate students rank the module in the top 20% of modules, and it also receives excellent feedback from the students participating in the student-staff liaison meetings organized by the department. But this is feedback for the complete module, not only for the part focusing on formal methods for security. Obviously, there are always a number of students that enjoy this content a lot, as evidenced by individual oral feedback and their selection of B.Sc. thesis topics in the following year. We plan to run more systematic feedback questionnaires for the individual topics taught in the undergraduate module in the next academic year.

Still, we consider our approach, at least in the more controlled undergraduate setting, a success: firstly, our students are able to apply formal methods thinking in controlled scenarios and, secondly, it helps us in attracting students for B.Sc. thesis topics that are related to our research. Last but not least, some of the students that did a B.Sc. thesis on formal methods with us, joined our groups as PhD students. Note that the University of Exeter, as most universities in the UK, admits students with undergraduate degree equivalent to at least a UK 2:1 Honors to its PhD program. Hence, we can offer our undergraduate students a direct path from a B.Sc. in Computer Science into our PhD program.

Acknowledgements. We would like to thank Sebastian Mödersheim for his valuable feedback on earlier versions of this paper.

References

1. Altenhofen, M., Brucker, A.D.: Practical issues with formal specifications. In: Kowalewski, S., Roveri, M. (eds.) FMICS 2010. LNCS, vol. 6371, pp. 17–32. Springer, Heidelberg (2010). https://doi.org/10.1007/978-3-642-15898-8_2
2. Backes, J., et al.: Semantic-based automated reasoning for AWS Access Policies using SMT. In: 2018 Formal Methods in Computer Aided Design (FMCAD), pp. 1–9 (2018). https://doi.org/10.23919/FMCAD.2018.8602994
3. Basin, D., Mödersheim, S., Viganò, L.: OFMC: a symbolic model checker for security protocols. Int. J. Inform. Secur. 4(3), 181–208 (2004). https://doi.org/10.1007/s10207-004-0055-7
4. Broy, M., et al.: Does Every Computer Scientist Need to Know Formal Methods? Formal Aspects of Computing (FAC) (2024)
5. Brucker, A.D., Stell, A.: Verifying feedforward neural networks for classification in Isabelle/HOL. In: Formal Methods (FM 2023). Ed. by Chechik, M., Katoen, J.-P., Leucker, M. Springer-Verlag (2023). https://doi.org/10.1007/978-3-031-27481-7_24
6. Dolev, D., Yao, A.: On the security of public key protocols. Sympos. Found. Comput. Sci. 0, 350–357 (1981). https://doi.org/10.1109/SFCS.1981.32
7. Dongol, B., et al.: On formal methods thinking in computer science education. Form. Asp. Comput. (2024). https://doi.org/10.1145/3670419
8. Greenaway, D., Andronick, J., Klein, G.: Bridging the gap: automatic verified abstraction of C. In: Beringer, L., Felty, A. (eds.) ITP 2012. LNCS, vol. 7406, pp. 99–115. Springer, Heidelberg (2012). https://doi.org/10.1007/978-3-642-32347-8_8
9. Hess, A.V., Mödersheim, S., Brucker, A.D., Schlichtkrull, A.: Automated Stateful Protocol Verification. Archive of Formal Proofs (2020)
10. Hess, A.V., Mödersheim, S., Brucker, A.D., Schlichtkrull, A.: Performing security proofs of stateful protocols. In: Computer Security Foundations Symposium (CSF), pp. 143–158. IEEE (2021). https://doi.org/10.1109/CSF51468.2021.00006
11. Lowe, G.: An attack on the Needham-Schroeder public-key authentication protocol. Inf. Process. Lett. 56(3), 131–133 (1995). https://doi.org/10.1016/0020-0190(95)00144-2
12. Mödersheim, S.: Protocol Security Verification Tutorial. Tech. rep., (2018)
13. Moore, J., Heule, M.: Industrial use of ACL2: applications, achievements, challenges, and directions. In: Reger, G., Traytel, D., Workshop on Automated Reasoning: Challenges, Applications, Directions, Exemplary Achievements (ARCADE). EPiC Series in Computing, pp. 42–45. EasyChair (2017). https://doi.org/10.29007/dh3f
14. de Moura, L., Kong, S., Avigad, J., van Doorn, F., von Raumer, J.: The lean theorem prover (System Description). In: Felty, A.P., Middeldorp, A. (eds.) CADE 2015. LNCS (LNAI), vol. 9195, pp. 378–388. Springer, Cham (2015). https://doi.org/10.1007/978-3-319-21401-6_26
15. Needham, R.M., Schroeder, M.D.: Using encryption for authentication in large networks of computers. Commun. ACM 21, 993–999 (1978). https://doi.org/10.1145/359657.359659
16. Paulson, L.C.: Inductive analysis of the internet protocol TLS. ACM Trans. Inf. Syst. Secur. 2(3), 332–351 (1999). https://doi.org/10.1145/322510.322530
17. Paulson, L.C., Nipkow, T., Wenzel, M.: From LCF to Isabelle/HOL. Formal Aspectsf Comput. (FAC) 31(6), 675–698 (2019). https://doi.org/10.1007/s00165-019-00492-1

Author Index